Felipe Carlos Yon Torres

Timing for outcrossing

Felipe Carlos Yon Torres

Timing for outcrossing

Circadian Clock Regulates Floral Rhythms with Large Fitness Consequences

Südwestdeutscher Verlag für Hochschulschriften

Impressum / Imprint
Bibliografische Information der Deutschen Nationalbibliothek: Die Deutsche Nationalbibliothek verzeichnet diese Publikation in der Deutschen Nationalbibliografie; detaillierte bibliografische Daten sind im Internet über http://dnb.d-nb.de abrufbar.
Alle in diesem Buch genannten Marken und Produktnamen unterliegen warenzeichen-, marken- oder patentrechtlichem Schutz bzw. sind Warenzeichen oder eingetragene Warenzeichen der jeweiligen Inhaber. Die Wiedergabe von Marken, Produktnamen, Gebrauchsnamen, Handelsnamen, Warenbezeichnungen u.s.w. in diesem Werk berechtigt auch ohne besondere Kennzeichnung nicht zu der Annahme, dass solche Namen im Sinne der Warenzeichen- und Markenschutzgesetzgebung als frei zu betrachten wären und daher von jedermann benutzt werden dürften.

Bibliographic information published by the Deutsche Nationalbibliothek: The Deutsche Nationalbibliothek lists this publication in the Deutsche Nationalbibliografie; detailed bibliographic data are available in the Internet at http://dnb.d-nb.de.
Any brand names and product names mentioned in this book are subject to trademark, brand or patent protection and are trademarks or registered trademarks of their respective holders. The use of brand names, product names, common names, trade names, product descriptions etc. even without a particular marking in this works is in no way to be construed to mean that such names may be regarded as unrestricted in respect of trademark and brand protection legislation and could thus be used by anyone.

Coverbild / Cover image: www.ingimage.com

Verlag / Publisher:
Südwestdeutscher Verlag für Hochschulschriften
ist ein Imprint der / is a trademark of
OmniScriptum GmbH & Co. KG
Heinrich-Böcking-Str. 6-8, 66121 Saarbrücken, Deutschland / Germany
Email: info@svh-verlag.de

Herstellung: siehe letzte Seite /
Printed at: see last page
ISBN: 978-3-8381-3917-3

Zugl. / Approved by: Jena, Friedrich-Schiller-Universität, Diss., 2014

Copyright © 2014 OmniScriptum GmbH & Co. KG
Alle Rechte vorbehalten. / All rights reserved. Saarbrücken 2014

Table of Contents

Chapter 1 – General Introduction 3
Chapter 2 - Manuscript Overview 15
Chapter 3 – Manuscript I 19
Chapter 4 – Manuscript II 50
Chapter 5 – Manuscript III 82
Chapter 6 – Discussion 102
Summary 116
Zusammenfassung 118
Bibliography 120
Acknowledgements 134

Chapter 1 – General Introduction

Angiosperm plants that depend on an animal-pollination strategy must advertise themselves to lure pollinators into their flowers. As a result, animal-pollinated plants have developed several floral traits, which can be divided mainly into visual traits, scent emissions and time-synchronization. Different floral traits on each category can be combined and grouped in response to different pollination syndromes, for example plants depending on medium-large size nocturnal-pollinators like moths and bats make a pollination syndrome group, or plants relying on diurnal small pollen feeders make another group. The right display of the floral traits often matches the activity period of the most efficient pollinators, for which the plant might have evolved or ecologically adapted to maximize its reproductive fitness. Here to make a distinction a plastic trait can be an ecological adaptation in a localized range compared to a general evolved trait for all the habitat range. Not all pollinators are equally efficient for a floral trait set, even when the traits have developed under similar selection pressures. For instance the case of hawkmoths and hummingbirds which share a similar foraging behavior thus making the plants develop excluding strategies that ensure the best fitness. In the case of these two different pollinators a strategy would be to develop a temporal partition, important to separate both groups with different efficiency, because Sphingid hawkmoths are in general nocturnal pollinators, compare to hummingbirds.

The results presented for this doctoral thesis are centered on the floral traits regulated by the circadian clock to maximize its plant-pollinator interaction. The thesis is divided in three manuscripts, taking the coyote tobacco *Nicotiana attenuata* as a plant model. First, I addressed the identification and conserved functionality of the circadian clock genes in *N. attenuata*, including its implication in flowering time. Second, I characterized the floral traits: corolla limb aperture, vertical movement and benzyl acetone emission on wild type (WT) plants, and empty vector (EV) as control for the transgenic clock-silenced lines. Additionally I provided evidence supporting a differential clock regulation between vegetative and floral tissue. Using the results of previous manipulation of the upward night flower orientation in *N. attenuata* to demonstrate its importance for interacting with the hawkmoth, *Manduca sexta*, and the existing knowledge on the other two traits, as third part I investigated how different clock components alter floral traits' display and time in order to assess their pollination fitness in glasshouse and field conditions.

Flower adaptation to pollinators

Floral traits such as reward and advertisement have been developed through evolutionary time in the Angiosperms to increase their chance of reproductive success (Grant 1949) by luring animal pollinators that transfer pollen between flowers. Prior to these developments, Gymnosperms depended on wind and water to transport their pollen (Harder & Barrett 2006), same as several present Angiosperms groups such as grasses (Poaceae).

The appearance of these floral traits have to precede any interaction with the pollinators (Stebbins 1970; Hodges & Arnold 1994, 1995), as a selective pressure cannot exist without an organ on which to exert it. From this point we can define the aleatory mutation as the starting point that recalls the attention of a pollinator which develops behavioral (deception) and/or physiological (nutrition) mutualism with flowers (Fenster *et al.* 2004). The development of flower traits also responds to avoid nectar and pollen robbers, and/or florivores (Harder & Barrett 2006)that feed on floral tissues, fruits or seeds, decreasing the reproductive fitness. In some cases pollinators can act as herbivores, not by consuming the flower directly but through their offspring, *i.e.* larvae, which consume plant tissues (Roda *et al.* 2004).

The evolution of floral traits may not be seen under current pollinator interactions, because some of the flower specializations might have evolved under extinct species and their interactions. These specializations might have nowadays readapted to extant species. This can be exemplified with invading plant species, where specialized floral traits fit or repurpose depending on available pollinators (Aizen *et al.* 2008). A flower is constituted by several different tissues, such as corolla, sepal, anthers, pistil, etc. which can evolve differentially in response to different pollinators if none of these exert a negative influence over the other pollinators (Hurlbert *et al.* 1996).

Pollination syndrome

Floral traits grouped by pollinator functional groups are defined as pollination syndromes, when the traits are selected to improve the efficiency of outcrossing by specific floral sizes and shapes according to the pollinator physiology and behavior (Fenster *et al.* 2004). Most general cases focus on rewarded interactions, and can be separated between those feeding on pollen and/or nectar, or, in special cases, oil rewards (Buchmann 1987). The interest of this research is to focus on one pollinator functional group, the one composed by Sphingid hawkmoths and hummingbirds because these two exert similar selection pressures over the floral traits (Grant 1952; Hodges *et al.* 2004). Flowers favored by these animals present long tubular corollas that allow an easier access of the proboscis/tongue to the nectar at the bottom of the corolla or either to a nectar spur (Hodges & Arnold 1995). Flowers present flat wide corolla limbs, either zygomorphic or actinomorphic. The coloration depends if it is specialized on hummingbirds, having then a red color, or white if on hawkmoths. Pattern colorations are of importance for the later, where grooves and black visual guides are more appealing while darker colorations tend to be ignored (Goyret 2010).

The flower orientation plays another important role, as the flying behavior of hummingbird and hawkmoths are different, giving a different visual perspective. Hummingbirds have a more flying free style and tend to move between the plants, allowing them to be more able to recognize flowers at different orientations (Fulton & Hodges 1999). On the contrary, hawkmoths have a hovering flight, observing the flowers from above (Sprayberry & Daniel 2007; Sprayberry & Suver 2011). At least in the case of *Hyles lineata*, the capability of recognizing flowers from which to feed is reduced mainly to those on upright orientation(Fulton & Hodges 1999; Hodges *et al.* 2004). The flower orientation is addressed on Manuscript III, where the flower angle is altered in clock-silenced lines, and their pollination relevance tested.

Natural history of the circadian clock

The circadian clock is a molecular oscillator entrained by the planet's 24 h rotation. It has a vital role in synchronizing the daily performance of organisms through light and temperature cycles, allowing a compartmentalization of activities to the correct day-time (Sanchez *et al.* 2011; McClung 2013). For example, the internal clock in plants influences processes through all the development stages, such as hypocotyl growth, cotyledon movement, leaf movement, bolting and flowering, and also abiotic and biotic stress resistance (Yakir *et al.* 2007; Roden & Ingle 2009). It is of equal importance for short-medium life-span plants that require a fast seed set, and for perennials that need years to reach reproductive maturity, in some cases overwintering under freezing temperatures (Ramos *et al.* 2005).

The interconnected molecular processes governing circadian rhythms in several organisms from different taxa have been elucidated, for instance in *Drosophila melanogaster, Neurospora crassa, Synechoccocus elongatus,* and *Mus musculus*(Panda *et al.* 2002; Golden & Canales 2003; Doherty & Kay 2010). In the plant kingdom, the Brassicaceae *Arabidopsis thaliana* was initially chosen as a model plant for extensive investigation of the circadian clock, on which a core oscillator based on an extended three-loop model was proposed (Ueda 2006; Pokhilko *et al.* 2012).

The core oscillator in *A. thaliana* is composed of four genes, two expressing at dawn: LATE ELONGATED HYPOCOTYL (LHY) and CIRCADIAN CLOCK ASSOCIATED 1 (CCA1); the other two expressing at subjective dusk: TIMING OF CAB EXPRESSION 1/PSEUDO-RESPONSE REGULATOR 1 (TOC1/PRR1) and ZEITLUPE (ZTL) (Pokhilko *et al.* 2012). The core oscillator comprises the first loop, it forms a negative feedback loop where CCA1 and LHY bind to TOC1 promoter to repress its transcription and, similarly, TOC1 represses LHY/CCA1 transcription (Huang *et al.* 2012; Pokhilko *et al.* 2013). The ZTL regulates TOC1 through its protein turnover during night. ZTL proteins contain an F-box domain which is part of a Skp/Cullin/F-box (SCF)E3 ubiquitin ligase complex that marks the TOC1 proteins for ubiquitination and subsequent degradation, therefore decreasing TOC1 protein concentration overnight (Más *et al.* 2003). In the same way other proteins from the TOC1 family are also targeted (Harmer 2009), leaving an still open question of how many proteins can be targeted by this complex. Recent works from the last decade have also found some, if not all, of these circadian clock genes in varied plant species including rice, soybean, maize, and poplar (Murakami *et al.* 2007; Liu *et al.* 2009; Takata *et al.* 2009; Wang *et al.* 2011b), showing how conserved in eudicots and monocots (Izawa *et al.* 2002; Ramos *et* al. 2005; Kaldis & Prombona 2006; Miwa *et al.* 2006) this endogenous clock is, playing therefore a central regulation role.

The second loop is formed by TOC1 and three genes: EARLY FLOWERING 3 (ELF3), ELF4, and LUX ARRHYTHMO (LUX) which are designated as Evening Complex (EC), with a peak expression at subjective dusk (Covington *et al.* 2001; Kikis *et al.* 2005; Helfer *et al.* 2011; Nusinow *et al.* 2011). The third loop is formed by another negative feedback loop between PRR7/PRR9 and LHY/CCA1, where LHY/CCA1 activate PRR7/PRR9 transcription and repress the EC. EC can inhibit itself and also represses PRR7/PRR9. This last duo represses LHY/CCA1 transcription instead (Zeilinger *et al.* 2006).

The circadian clock requires external signals for entrainment, which provides it with the possibility to adapt to different photoperiod and thermoperiod (Harmer 2009). This has been extensively studied in laboratory conditions, which allowed to identify the main regulatory agents due to its stable conditions. Some studies have shown the importance of the circadian clock to make the most of its resources through nutrient assimilation and photosynthesis in order to maximize its fitness (Roenneberg 1994; Green *et al.* 2002; Dodd *et al.* 2005). Nevertheless, light and temperature oscillate in natural conditions in a greater range without perfect repetitive patterns over days. A series of other interactions also provide extra input for clock regulation, such as drought and herbivory. Few works have been conducted on field conditions to study the same clock principles (Resco *et al.* 2009), as demonstrated on rice studies in Japan, where mutant phenotypes were rescued by the strong environmental signals in the field, which entrained the endogenous clock to its normal oscillation (Izawa 2012; Nagano *et al.* 2012).

Circadian clock regulation of floral rhythms

Currently most studies between circadian clock and flowers have been limited to the flowering time, the flower aperture, and scent emission. Many of these observations were done indirectly for testing non circadian hypothesis, and none of those studies used circadian clock mutants. The timing of elongation and flowering has been widely studied in several species, as it differs between the location and weather conditions that a plant inhabits (Fournier-Level *et al.* 2011). In the case of *Arabidopsis*, a long day flowering plant, it requires the photoperiod to reach a critical length to match the light signaling and circadian molecular mechanism that triggers the change to reproductive stage (Sawa *et al.* 2007). Homologous genes of ZTL are involved in the circadian regulation of photoperiodic flowering. This gene family known as ADAGIO involves FLAVIN-BINDING, KELCH REPEAT, F-BOX 1/ADAGIO3 (FKF1/ADO3) that interacts with GIGANTEA (GI) in a blue light-dependent manner and regulates CONSTANS (CO) expression by degrading the CYCLING DOF FACTOR 1 (Imaizumi & Kay 2006). Additionally, the third ADAGIO gene member LOV KELCH PROTEIN 2 (LOV2/ADO2) works together with ZTL and FKF1 to regulate the protein degradation of TOC1 and PRR5. These last two interact with ELF3/ELF4/LUX contributing to the circadian oscillation (Baudry *et al.* 2010).

Given the variety of habitats, plants respond in different ways to the different photoperiod lengths that trigger flowering, being possible to classify plants as short, neutral or long day photoperiod. Opposite to *Arabidopsis*, rice requires a short day to flower matching with the autumn's start in the northern hemisphere (Itoh & Izawa 2013). Also other species are dissimilar, for which the standard *Arabidopsis* model can't be applied as a general rule, neither the expected effects of their mutants can be inferred for all.

In the past, the flower aperture was observed and described in the works of Linnaeus (Linnaeus 1751) who observed the distinctive flower opening time of different species, inferring that flower aperture is under an endogenous clock control. Nonetheless, studies using knock-out circadian clock genes, either mutants or by silencing, have not been made (van Doorn & Van Meeteren 2003; Resco *et al.* 2009).

Given the variety of circadian rhythms observed within a same plant, the question of a clock multiplicity had arisen because different processes require different synchronization, like leaf movement, photosynthesis, stomata conductance, catalase activity, etc. (Harmer 2009). Currently, there is even evidence of a circadian clock driven entirely by reactive oxygen species (Edgar *et al.* 2012) or sugar metabolism (Stitt & Zeeman 2012). In consideration of these examples, it is worthwhile to ask if reproductive tissues can also have a different clock regulation or even one of their own, a question that is addressed on Manuscript II, combining molecular and phenotypic data.

Nicotiana attenuata as a clock-ecological interaction model

In this investigation, the Solanaceous plant *Nicotiana attenuata*, commonly known as coyote tobacco, was employed to study the effects of the circadian clock regulation. Ecological research has been previously done on areas such as abiotic response(Dinh *et al.* 2013), herbivore's interaction(Wu *et al.* 2008; Schäfer *et al.* 2011), defense metabolism (Voelckel *et al.* 2001; Wu *et al.* 2007; Steppuhn *et al.* 2008), and pollinator's interaction (Kessler & Baldwin 2007; Kessler 2012). The natural habitat of *N. attenuata* is the Great Basin Desert in the United States, where it inhabits a large range of landscapes (Goodspeed 1954) with their particular meteorological conditions, abiotic and biotic factors, and adaptations(Zavala *et al.* 2004). *N. attenuata* is a seasonal plant, which completes its life cycle during spring and summer to lie dormant in the seed bank for the many years between fires in its native habitat (Baldwin *et al.* 1994).

The flowers of *N. attenuata* have visual and olfactive traits. It has been previously described as a flower with two distinctive flower opening times: the first type opens at night (after dusk) until early morning and comprises 90% of the total, and the second opens in the morning between 6 - 8 h and only to 50% of the full corolla limb extent. Both flower types close normally after 8 h until they reopen later in the day as normal night flowers for a second and third time. Together with the flower opening, there is the emission of a floral bouquet, on which benzyl acetone (BA) is the main component. BA is mainly emitted at night, for which morning flowers have little scent emission; this is because scent emission intensity is associated with the corolla limb expansion, so less BA is emitted during a partial opening (Kessler *et al.* 2010).

A third trait is *N. attenuata*'s flower vertical movement, described for the first time in WT and also characterized in clock-silenced lines in Manuscript II. This vertical movement falls within visual traits, as it makes the corolla limb more conspicuous depending on the flower orientation, which goes on WT approximately from -90° to 40°. It is considered a clock regulated trait, because its movement follows a rhythmic pattern during the flower life span, up at night and down during the day.

The pollinator dilemma

It is usually considered that a plant species specialize only into a pollinator functional-group(Stebbins 1970) and within this group to particular set of species(Grant 1949), as is the case of *Nicotiana attenuata* with *Manduca* spp. hawkmoths. As the flower shape and size are characteristic of big pollinator's functional group, this specialization can be considered a temporal one. The raised question through several plant-pollinator interactions is: when does the plant should specialize in a single pollinator? The mainstream theory proposed by Stebbins in the '70s, "Most Effective Pollinator Principle" (MEPP), implies that a plant will always evolve specializations to maximize its interaction with the most effective pollinator. Currently this theory is controverted because it has been found that assemblages of pollinator generalists work at unison with pollinator specialists, and several of the well-known examples of specialized flowers are dependent on pollinator generalists (Waser *et al.* 1996). As demonstrated by the study of Hurlbert (Hurlbert *et al.* 1996) in *Impatiens capensis*, a plant does not necessarily need to adapt to a single pollinator type if it does not bring considerable negative effects on the other pollinators. It is the case of hummingbirds and bees, where the development of spurs and wider corolla limb does not interfere with the foraging behavior of any of both respectively.

It is important for this research to describe the animal pollinator interaction of *N. attenuata*. There are multiple pollinators in the Great Basin Desert, which can be divided as night active or day active. The most important night-time pollinators are hawkmoths from the Sphingidae family, as so far agreed. In this habitat the hawkmoth pollinators are composed by *Manduca sexta* and *Manduca quinquemaculata*, and *Hyles lineata*. Both *Manduca* spp. are considered the most effective pollinators of *N. attenuata*, having their activity peak after dusk, around 22 h. Instead, the active time of *H. lineata*, comprises the time before and after dusk. This whole pollinator group feeds on nectar, for which they can repeatedly visit the same flower over the nights of the flower's life span.

In the day time we can find a variety of pollinator groups, among hummingbirds and mainly Hymenoptera insects. The principal day pollinator is the black-chinned hummingbird, *Archilochus alexandri*, which preferentially pollinates morning flowers (Kessler *et al.* 2010). In the case of Hymenoptera, it has been observed that bees and sweat bees visit *N. attenuata* flowers, but these two groups are mainly pollen feeders, which will visit the flowers mostly on the first day of opening, as later pollen will be depleted. In pollination experiments of Manuscript III, bees were not considered as the target flowers were emasculated. In contrast, hummingbirds visit flowers that contain nectar, so they can overtime learn to choose the most nutritious flower in an assemblage of a given plant species.

As both pollinators of *N. attenuata*, hawkmoths and hummingbirds, belong to the same functional group, in Manuscript III we tested how this floral temporal regulation is controlled by single clock genes and its effect on the reproduction fitness mediated by night- and day-time pollinators.

Chapter 2 - Manuscript Overview

Manuscript I

Identification and characterization of circadian clock genes in a native tobacco, *Nicotiana attenuata*.

Felipe Yon, Pil Joon Seo, Jae Yong Ryu, Chung-Mo Park, Ian T. Baldwin and Sang-Gyu Kim

In this manuscript, we identified the homologue core clock genes in *Nicotiana attenuata* of *Arabidopsis* clock genes through the protein similarity and time specific transcript accumulation. By performing the transformation of NaLHY and NaZTL in *Arabidopsis* and protein interaction assays, we confirmed the functional conservation of our homologous clock genes. Additionally, we found that silencing NaTOC1 in *N. attenuata* results in late flowering, supporting that the circadian clock in N. attenuata regulates flowering time as *Arabidopsis* clocks do.

Manuscript II

Nicotiana attenuata LHY and ZTL regulate circadian rhythms in flowers

Felipe Yon, Youngsung Joo, Eva Rothe, Ian T. Baldwin, Sang-Gyu Kim

In this manuscript, we examined the expression of NaLHY and NaZTL transcripts among flower tissues and the regulation of floral traits through the silencing of these genes in *N. attenuata*. We examined the differential time shift of clock genes expression between floral tissue and non-floral tissue. We demonstrated the importance of single clock components in the time regulation and functional display of floral traits, such as flower vertical movement, flower open and close timing, and the emission of its main floral volatile, benzyl acetone.

Manuscript III

Fitness consequences of altering circadian rhythms in *Nicotiana attenuata* flowers

Felipe Yon, Danny Kessler, Lucas Cortés Llorca, Youngsung Joo, Eva Rothe, Ian T. Baldwin and Sang-Gyu Kim

In this manuscript, we demonstrated the ecological implications of an altered clock in the floral traits of *Nicotiana attenuata*. Specifically how these alterations change the fitness outcome under glasshouse and natural field conditions, when pollination is mediated by nocturnal *Manduca sexta* hawkmoth and diurnal hummingbirds. We evidence that a fully functional clock is not necessarily advantageous given a certain environmental context, as its flexibility should allow for local adaptations.

Chapter 2 – Manuscript Overview

Chapter 3 – Manuscript I

BMC Plant Biology 2012, **12**:172 doi:10.1186/1471-2229-12-172

Identification and characterization of circadian clock genes in a native tobacco, *Nicotiana attenuata*

Felipe Yon [1†], Pil-Joon Seo [2,3†], Jae Yong Ryu [2], Chung-Mo Park [2], Ian T Baldwin [1] and Sang-Gyu Kim [1]

[1] Department of Molecular Ecology, Max Planck Institute for Chemical Ecology, Hans-Knöll-Straße 8, Jena D-07745, Germany

[2] Department of Chemistry, Seoul National University, Seoul, 151-742, Korea

[3] Department of Chemistry, Chonbuk National University, Jeonju, 561-756, Korea

† Equal contributors

Abstract

Background: A plant's endogenous clock (circadian clock) entrains physiological processes to light/dark and temperature cycles. Forward and reverse genetic approaches in *Arabidopsis* have revealed the mechanisms of the circadian clock and its components in the genome. Similar approaches have been used to characterize conserved clock elements in several plant species. A wild tobacco, *Nicotiana attenuata* has been studied extensively to understand responses to biotic or abiotic stress in the glasshouse and also in their native habitat. During two decades of field experiment, we observed several diurnal rhythmic traits of *N. attenuata* in nature. To expand our knowledge of circadian clock function into the entrainment of traits important for ecological processes, we here report three core clock components in *N. attenuata*.

Results: Protein similarity and transcript accumulation allowed us to isolate orthologous genes of the core circadian clock components, LATE ELONGATED HYPOCOTYL (LHY), TIMING OF CAB EXPRESSION 1/PSEUDO-RESPONSE REGULATOR 1 (TOC1/PRR1), and ZEITLUPE (ZTL). Transcript accumulation of *NaLHY* peaked at dawn and *NaTOC1* peaked at dusk in plants grown under long day conditions. Ectopic expression of *NaLHY* and *NaZTL* in *Arabidopsis* resulted in elongated hypocotyl and late-flowering phenotypes. Protein interactions between NaTOC1 and NaZTL were confirmed by yeast two-hybrid assays. Finally, when *NaTOC1* was silenced in *N. attenuata,* late-flowering phenotypes under long day conditions were clearly observed.

Conclusions: We identified three core circadian clock genes in *N. attenuata* and demonstrated the functional and biochemical conservation of *NaLHY, NaTOC1,* and *NaZTL.*

Background

The circadian clock, entrained by our planet's 24 h rotation on its tilted axis, plays crucial roles in the synchronization of the performance of organisms with daily cycles of light and temperature, enabling organisms to regulate activities at the correct time of a day [1]. For instance, the endogenous clock in plants influences various biological processes including leaf movements, hypocotyl growth, floral transition, and abiotic and biotic stress resistance [2-4].

The circadian rhythmicity and molecular mechanisms underlying the circadian clock have been investigated in many organisms including *Drosophila melanogaster*, *Neurospora crassa*, *Synechoccocus elongatus*, and mice [5-7]. In general, several interconnected transcription/translation feedback loops participate to establish central clock oscillations [8-10]. In plants, circadian rhythmicity is extensively investigated in a dicotyledonous model plant, *Arabidopsis thaliana*, and a 'three-loop model' has been proposed [11].

TIMING OF CAB EXPRESSION 1/PSEUDO-RESPONSE REGULATOR 1 (TOC1/PRR1) and two partially redundant MYB transcription factors, CIRCADIAN CLOCK-ASSOCIATED 1 (CCA1) and LATE ELONGATED HYPOCOTYL (LHY) comprise a central oscillation loop [12]. CCA1 and LHY repress transcript expression of *TOC1* by binding directly to its promoter, and as shown recently TOC1 negatively regulates the transcription of *CCA1* and *LHY* [13,14], establishing negative feedback loop [11,15]. The second loop is formed by TOC1 and the evening complex (EC) consisting of EARLY FLOWERING 3 (ELF3), ELF4, and LUX ARRHYTHMO (LUX) [14]. In addition, the third loop is established by negative feedback between PRR7/PRR9 and LHY/CCA1 [16,17].

Protein turnover regulation provides another layer of sophistication in the regulation of the circadian clock. Notably, the TOC1 protein is regulated by proteasome-mediated protein degradation. The F-box protein ZEITLUPE (ZTL) in the E3 ubiquitin ligase SCF (SKP1-CUL1-F-box protein) complex ubiquitinates the TOC1 protein through direct physical interaction in a dark-dependent manner [18]. Homologs of ZTL are also involved in circadian regulation of photoperiodic flowering [19,20]. The FLAVIN-BINDING, KELCH REPEAT, F-BOX 1/ADAGIO3 (FKF1/ADO3) interacts with GI in a blue light-dependent manner and regulates *CONSTANS* expression by degrading a Dof transcription factor, CYCLING DOF FACTOR 1 [19]. In addition, it has been reported that ZTL, FKF1, and LOV KELCH PROTEIN 2 together regulate TOC1 and PRR5 degradation, contributing to the circadian oscillation [21].

Recent works has shown that many of these circadian clock components can also be found in diverse plant species including rice, soybean, maize, and poplar [22-25]. *CCA1/LHY* genes are widely conserved in eudicotyledonous (eudicots) and monocotyledonous (monocots) plants [26-29], and CCA1/LHY and TOC1 feedback loops are thought to play a central role in the clock's function in these plant species. Functional homologs of ZTL have also been found in several plant species [30,31], further supporting that circadian clock components are fairly well-conserved in plants.

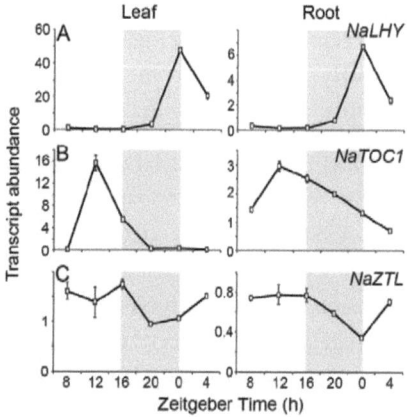

Figure 1 Diurnal rhythms of putative circadian clock genes in *Nicotiana attenuata*. Expression of orthologs of the circadian clock genes were first examined using our time course microarray data (accession number GSE30287). Wild-type N. attenuata plants were harvested every 4 h for one day from leaves and roots. After RNA isolation, each sample was hybridized on Agilent single color technology arrays designed from the *N. attenuata* transcriptome (accession number GPL13527). Mean (± SE) levels of transcript abundance of (A) NaLHY, (B) NaTOC1, and (C) NaZTL in leaves (n = 3) and roots (n = 3) at each harvest time. Gray boxes depict the dark period of LD (16 h light/ 8 h dark). Na, *Nicotiana attenuata*; LHY, late elongated hypocotyl; TOC1, timing of cab expression 1; ZTL, zeitlupe.

Despite the important role of the endogenous clock in entraining physiological processes to environmental signals, how the circadian clock regulates ecological performance of a plant in its natural habitat is largely unknown. Only a few studies have shown that the endogenous clock allows plants to maximize photosynthetic capacity and reproductive success [32-34]. In order to expand our understanding of the clock function in biotic and abiotic interactions, we

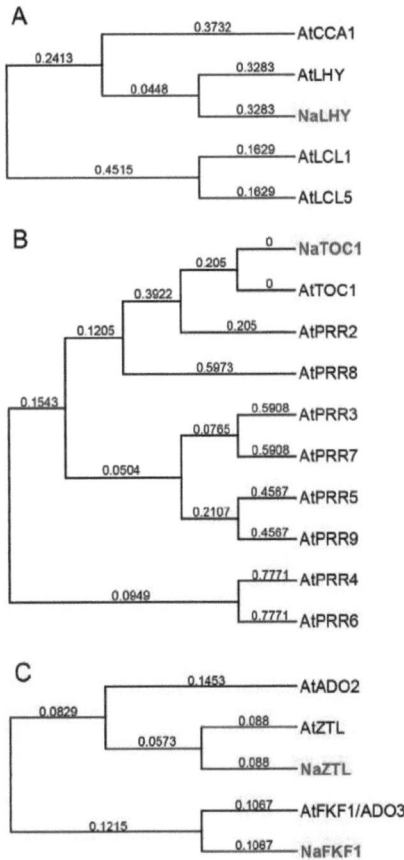

Figure 2 Phylogenetic trees of putative circadian clock genes in *N. attenuata*. Phylogenetic relationships among predicted orthologs of (A) LHY and CCA1, (B) TOC1 and PRRs, (C) ZTL and FKF1/ADO3 in *Arabidopsis thaliana* (At) and *N. attenuata* (Na). Full-length amino acid sequences were aligned and phylogenetic trees were reconstructed by the UPGMA method. The numbers given for each branch represent the numbers of amino acid substitutions per site. LHY, late elongated hypocotyl; CCA1, circadian clock-associated 1; TOC1, timing of cab expression 1; PRR, pseudo-response regulator; ZTL, zeitlupe; FKF1, flavin-binding kelch repeat F-box 1; ADO, adagio.

identified three core clock components (LHY, TOC1, and ZTL) in a wild tobacco, *Nicotiana attenuata*, which has been developed as a model system for understanding ecological performance in native habitats, in particular the Great Basin desert in Utah. *N. attenuata* is a seasonal solanaceous plant, completing its life cycle during spring and summer to lie dormant in the seed bank for the many years between fires in its native habitat. These results provide additional evidence of the conservation of the circadian clock genes and set the stage for future studies to unravel the ecological relevance of the clock.

Results

Isolation of putative core circadian clock genes in *N. attenuata*

The full-length or partial sequences of three putative core circadian clock genes (*LHY*, *TOC1*, and *ZTL*) in *N. attenuata* were isolated by BLAST search against in-house cDNA library using the sequences of *Arabidopsis* clock genes. We checked the diurnal expression of these transcripts in our time series microarray database [35], which examined patterns of transcript accumulation in *N. attenuata* leaf and root tissues every 4 h for one day (Figure 1). To confirm the microarray data and examine circadian rhythms of the selected clock genes in *N. attenuata*, we analyzed the transcript accumulation of the candidate genes in seedlings grown under different light conditions. *N. attenuata* plants were grown under 16 h light/ 8 h dark cycle (LD) for two weeks and subsequently transferred into continuous light condition (LL). Twenty seedlings in LD and LL were harvested every 2 h for three days. Quantification of mRNA expression was performed by quantitative real-time PCR (qPCR) using the gene specific primers (Additional file 1). We also constructed full-length of coding sequences of *NaLHY*, *NaTOC1*, and *NaZTL* (Additional file 2), based on the ortholog sequences available in public EST databases and *Arabidopsis* database. To examine evolutionary relationship of circadian clock genes in plant species, phylogenetic trees were constructed using the UPGMA algorithm (Figure 2 and Additional file 3).

Arabidopsis and poplar genomes contain two MYB transcription factors (LHY/CCA1 and LHY1/2, respectively), which play a key role in the regulation of the endogenous clock [23,36]. However, we found only one oscillating *AtLHY*-like transcript in our cDNA library (Figure 2A). The transcript levels of *NaLHY* in LD peaked at Zeitgeber time 0 h (ZT 0 h) and remained low until the next ZT 0 h (Figures 1A and 3A), consistent with the patterns observed in

Arabidopsis and rice [26,37]. *NaLHY* transcript accumulation under LL maintained a circadian rhythm peaking near subjective dawn and showed a shorter period of time with an average 6.1 h than under LD (Figure 3A). Protein sequence of NaLHY shared a relatively high similarity with AtLHY (Identities = 42%, Positives = 54%, Gaps = 18%) and AtCCA1 (Identities = 38%, Positives = 50%, Gaps = 22%). Phylogenetic analysis indicated that NaLHY was more closely related to AtLHY than to AtCCA1 (Figure 2). The NaLHY protein (767 aa) is larger than AtLHY (645 aa), AtCCA1 (608 aa), and OsCCA1 (719 aa) but has the conserved SANT domain at the N-terminal and the two alanine rich regions which characterize the AtLHY and OsCCA1 proteins (Additional file 2).

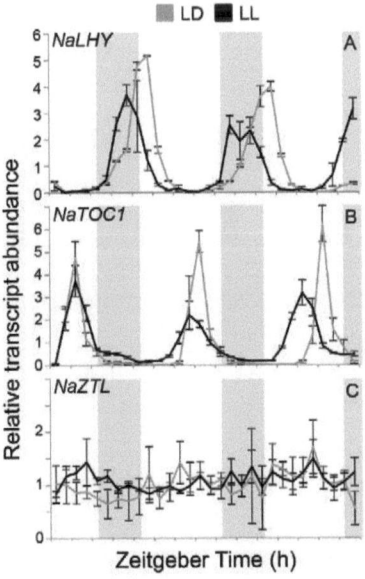

Figure 3 Circadian rhythm of the clock gene expression in *N. attenuata*. Mean (± SE) levels of relative transcript abundance of (A) NaLHY, (B) NaTOC1, and (C) NaZTL in wild-type N. attenuata seedling (n = 3) at each harvest time for two days under long day condition (LD, gray lines, 16 h light/ 8 h dark) and continuous light condition (LL, black line). Gray boxes depict the dark period of LD. A *N. attenuata* elongation factor gene was used as a control for constitute expression.

TOC1/PRR1 is a member of the PRR family, which is composed of five oscillating genes; *TOC1/PRR1*, *PRR3*, *PRR5*, *PRR7* and *PRR9*. Each of *PRR* genes has its own diurnal expression pattern [38]. The microarray data revealed that *NaTOC1* transcripts peaked at ZT 12 h (Figure 1B) in accordance with the known circadian rhythm of the *Arabidopsis TOC1* [39,40]. The qPCR analysis showed that *NaTOC1* transcripts peaked at ZT 12 h under LD and the expression of *NaTOC1* under LL peaked earlier by an average of 1.3 h compared to the expression under LD (Figure 3B). The full-length NaTOC1 protein sequence exhibited high similarity to AtTOC1/PRR1 (Identities = 49%, Positives = 59%, Gaps = 16%). Phylogenetic analysis revealed that NaTOC1 is most closely related to AtPRR1 than to other AtPRRs (Figure 2). The REC domain at the N-terminal and the CCT motif in the C-terminal of TOC1 were conserved in *N. attenuata*, *Arabidopsis* and *O. sativa* but the coiled-coil region was only found in the eudicots *Arabidopsis* and *N. attenuata* (Additional file 2).

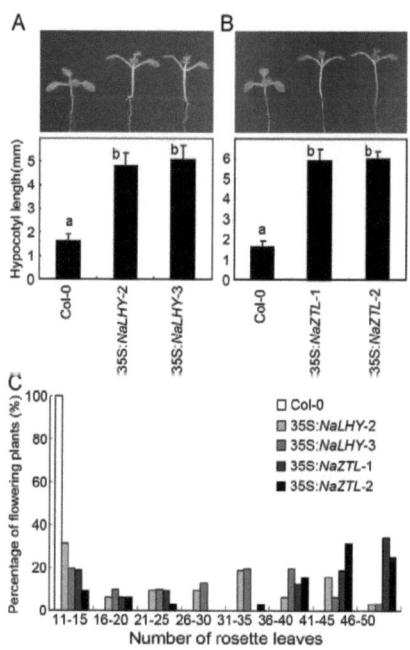

Figure 4 Ectopic expression of NaLHY and NaZTL in *A. thaliana*. Arabidopsis transgenic plants harboring 35S:NaLHY and 35S:NaZTL constructs are shown at seedling (A, B) and flowering (C) stages. Mean (± SE) values of hypocotyl lengths in wild-type Col-0 (n=11), 35S:NaLHY-2 (n=12), 35S:NaLHY-3 (n=11), 35S:NaZTL-1 (n=13), and 35S:NaZTL-2 (n=13). Different letters (a and b) reflect significant differences among the lines ($P<0.05$, one-way ANOVA with Fisher's post hoc test). (C) The percentage of flowering plants and number of rosette leaves of Col-0 (n=16), 35S:NaLHY-2 (n=32), 35S:NaLHY-3 (n=31), and 35S:NaZTL-1 (n=32), 35S:NaZTL-2 (n=30) when inflorescence elongation started.

The *Arabidopsis* genome encodes three F-box proteins involved in protein degradation of the clock components [18,19,21]. We found two *ZTL* orthologous genes (*NaZTL* and *NaFKF1/ADO3*) in our cDNA library and named them according to the phylogenetic analysis (Figure 2). *NaZTL* transcripts under LL and LD were largely unchanged (Figures 1C and 3C), consistent with the results from *Arabidopsis* [41]. In contrast, *NaFKF1* transcripts showed a clear circadian rhythm with peaks at ZT 12 h under LD (Additional file 4). The period of *NaFKF1* expression in LL was shortened by an average of 2 h (Additional file 4). NaZTL had protein sequence similarity to AtZTL (Identities = 79%, Positives = 86%, Gaps = 5%) and NaFKF1 with that of AtFKF1 (Identities = 78%, Positives = 85%, Gaps = 4%). NaZTL and NaFKF1 proteins contained LOV/PAS and F-box domains at their N-terminals and Kelch repeats region in their C-terminals as shown in orthologs in *Arabidopsis* (Additional file 2).

Ectopic expression of *N. attenuata* clock genes in *arabidopsis*

To examine the functional conservation of the clock components in *N. attenuata*, we produced *Arabidopsis* (Col-0) transgenic plants ectopically expressing *NaLHY* and *NaZTL* genes under the control of Cauliflower Mosaic Virus (CaMV) 35S promoter (35S:*NaLHY* and 35S:*NaZTL*). The circadian clock entrains hypocotyl cell elongation to light–dark cycles, and thus plants overexpressing either *Arabidopsis LHY* or *ZTL* have elongated hypocotyls compared to wild-type plants [37,42]. We measured the hypocotyl lengths of 35S:*NaLHY* (Figure 4A) and 35S:*NaZTL* transgenic plants (Figure 4B). Two independent 35S:*NaLHY* lines showed a pronounced increase in hypocotyl length compared to wild-type, Col-0, as did the two independent 35S:*NaZTL* lines ($P < 0.05$, one-way ANOVA with Fisher's *post hoc* test).

The ability to perceive seasonal changes by the circadian clock is required for the successful transition from vegetative to reproductive stages. Knocking-out a mor-ning element, *Arabidopsis* LHY or CCA1 results in an early-flowering phenotype and, in contrast, their overexpressing lines result in a late-flowering phenotype in *Arabidopsis* [12,36,37]. The overexpression of *Arabidopsis* ZTL also results in a late-flowering phenotype under LD [42]. To examine the effect of *NaLHY* and *NaZTL* on flowering time regulation in *Arabidopsis*, we scored the rosette leaf number at a time when initial flowering was observed in plants grown under LD. Both 35S:*NaLHY* and 35S:*NaZTL* plants exhibited delayed

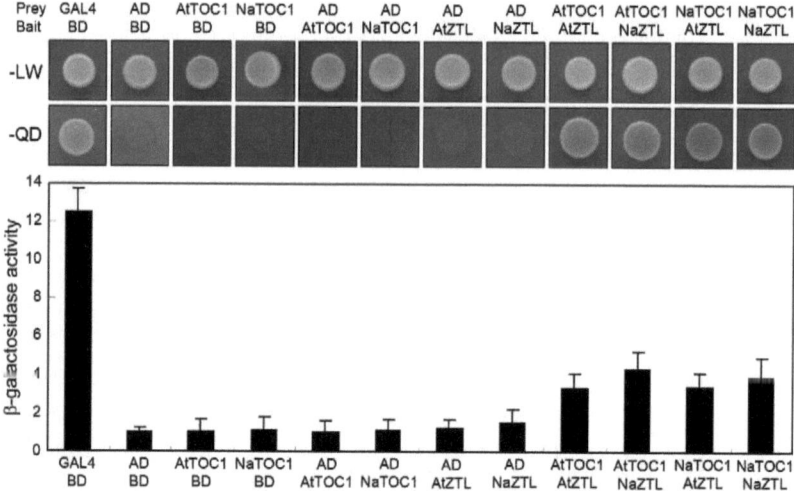

Figure 5 Protein interactions between TOC1 and ZTL from N. attenuata and A. thaliana using a yeast two-hybrid assay. (A) Growth of the yeast cells carrying prey and bait constructs indicated on the top of each panel. The ZTL proteins were fused to the GAL4 DNA binding domain (bait), and the TOC1 proteins were fused to the GAL4 activation domain (prey). The yeast cells can grow on QD medium when bait and prey proteins physically bind. (B) Mean (± SE) levels of β-galactosidase activity of the yeast carrying prey and bait constructs. BD, empty vector expressing binding domain; AD, empty vector expressing activation domain; -LW, synthetic dropout (SD) yeast growth medium lacking leucine and tryptophan; QD, SD medium lacking Leu, Trp, histidine, and adenine.

flowering phenotypes and increased rosette leaf numbers (Figure 4C).

Interaction between NaTOC1 and NaZTL protein

F-box protein ZTL plays a key role in protein turnover of the clock components, including TOC1 by direct protein-protein interactions [18]. We determined whether NaTOC1 interacts with NaZTL by yeast two-hybrid analysis. In addition, we tested inter-species interactions of AtTOC1-NaZTL and AtZTL-NaTOC1. Full-length NaTOC1 proteins (AtTOC1) fused with the GAL4 activation domain (AD), and NaZTL (AtZTL) fused with the GAL4 binding domain (BD) were co-expressed in yeast cells (see Methods). The yeast cells expressing both NaTOC1 and NaZTL constructs grew well in the absence of leucine, tryptophan, histidine, and adenine, showing the positive interaction, consistent with *Arabidopsis* TOC1 and ZTL interaction (Figure 5A). The transformants expressing both NaTOC1 and NaZTL also exhibited strong β-galactosidase activity compared to control yeast cells carrying empty vectors or only one of the constructs (Figure 5B). Inter-species protein interactions of AtTOC1-NaZTL and AtZTL-NaTOC1 were as strong as the intra-species interactions of TOC1 and ZTL (Figure 5).

Silencing of *NaTOC1* in *N. attenuata*

To examine the conserved and unique functions of circadian clock genes in *N. attenuata*, we first silenced *NaTOC1* expression by constitutive overexpression of *NaTOC1*-specific inverted repeat (ir) sequences (ir*TOC1* lines) [43,44]. TOC1 in *Arabidopsis* plays a key role in flowering time regulation. The semi-dominant *toc1-1* mutant displays a late-flowering phenotype under LD and an early-flowering phenotype under short day conditions (SD) [39] and the *toc1-2* loss-of-function mutant shows no significant difference of flowering time under LD but an early-flowering phenotype under SD [45]. During the screening of the silenced lines, we clearly observed the late-flowering phenotypes of several independent ir*TOC1* lines under LD (Figure 6). Two independent lines were used to measure flowering time and rosette leaf number when the first flower opened (Figure 6C). The silencing efficiency of independent lines harboring the RNAi construct correlated strongly with the flowe-ring phenotype (Figure 6B). In one homozygous ir*TOC1*-162 line, screened on the basis of antibiotic resistance, but in which the *TOC1* gene expression was not silenced, presumably due to methylation of the ir part of the T-DNA or an unknown insertion effect, displayed a WT flowering pattern (Figure 6).

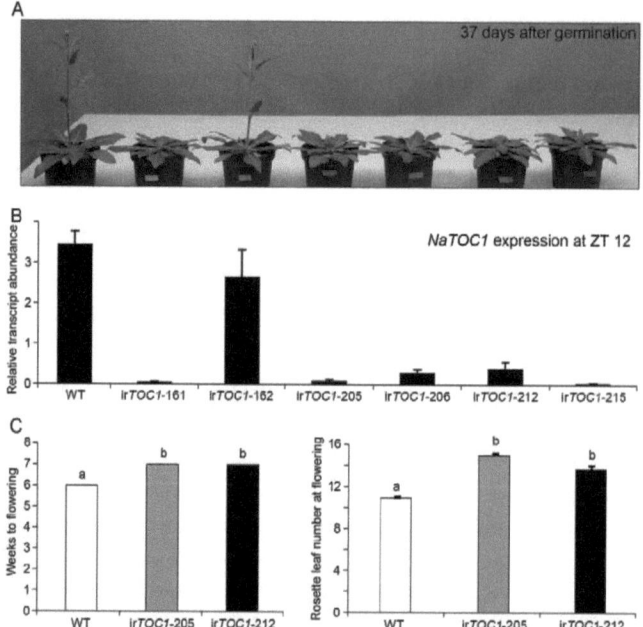

Figure 6 Late-flowering phenotypes of NaTOC1 silenced *N. attenuata*. (A) Phenotypes of transgenic plants silenced in NaTOC1 gene expression (irTOC1) with an inverted repeat (ir) RNAi construct. (B) Mean (± SE) levels of NaTOC1 (n=3) transcript accumulation at zeitgeber time (ZT) 12 in wild-type and in several irTOC1 lines. (C) Time on weeks of WT and 2 selected irTOC1 lines to reach flowering stage (n=5). (D) Mean (± SE) number of leaves at flowering stage (n=5).

Discussion

Since the clock genes of *Arabidopsis* have been identified, the clock mechanism of *Arabidopsis* has been extended to other dicotyledonous plants; soybean [24], chestnut [29], *Brassica rapa*[46], poplar [23] and also to monocotyledonous plants; rice [25], maize [22], duckweed [28]. Here, we identified three core circadian clock genes in a wild tobacco, *N. attenuata*. Analyses of the circadian rhythms of transcript accumulation and protein similarity allowed us to identify orthologs of *LHY*, *TOC1*, and *ZTL* of *Arabidopsis* in *N. attenuata*. Protein interactions of TOC1 and ZTL in *Arabidopsis* were also conserved in *N. attenuata*. In addition, ectopic expression of *NaLHY* and *NaZTL* in *Arabidopsis* confirmed the functional conservation of LHY and ZTL in *N. attenuata*. We also demonstrated that *NaTOC1* in *N. attenuata* plays an important role in the regulation of flowering time.

The number of *LHY* (or *CCA1*) orthologous genes and the timing of gene duplication events differ among plant species [23]. One common ancestor of LHY/CCA1 independently duplicated in monocots and eudicots. *Populus nigra* and *P. trichocarpa* contain two *LHY* orthologs and the monocots, rice and *Sorghum bicolor* contain one *CCA1*-like gene in their genome [23,47]. Gene duplication events of *LHY/CCA1* in popular and *Arabidopsis* would be expected in *N. attenuata* but we were able to find only one *LHY/CCA1* like gene in *N. attenuata*. However, the second ortholog could be missing in our current cDNA library and we plan to perform deep sequencing of the transcriptome and microarray analysis of plants grown under a variety of conditions in the future to clarify the evolutionary relationship of *N. attenuata*'s *LHY* orthologs.

While components of the endogenous clock and their associated circadian clock mechanisms have clearly been maintained across diverse plant species, clock-mediated signaling has evolved in response to differing selection pressures. A perennial plant, chestnut (*Castanea sativa*) has orthologs of *AtLHY* and *AtTOC1* in its genome and the pattern of transcript accumulation of *CsLHY* and *CsTOC1* is similar to that of *Arabidopsis* in LD at 22°C [29]. However, in winter condition, both *CsLHY* and *CsTOC1* transcripts lose their diurnal rhythms and maintain high levels of transcripts, which may be associated with the induction of winter dormancy [29]. Even within a single species, *A. thaliana*, genetic variation in the clock components plays a critical role in adapting ecotypes to their local environment [48]. Genetic variation in the *PRR* genes is associated with local adaptation seen in the differential expression of quantitative trait loci in 150 *Arabidopsis* ecotypes [48]. In addition, the degree to which TOC1 regulates flowering time differs among *Arabidopsis* ecotypes. A semi-dominant *toc1-1* mutant of the *Arabidopsis* C24 ecotype displays late a flowering phenotype in LD, whereas the same *TOC1* mutation in the Landsberg ecotype results in no change in flowering time compared with WT plants under LD [39]. The *toc1-2* mutant shows also no change in flowering time under LD [45]. However, silencing *TOC1* in *N. attenuata* confers a late-flowering phenotype under LD, which may be due to the longer life span of *N. attenuata* (about 3 months) or different circadian clock functions have evolved under various environmental pressures. In future research, we plan to measure the flowering time of ir*TOC1* lines under SD to examine the light sensitivity of this transgenic line.

We have investigated the ecology of *N. attenuata* in its native habitat for more than twenty years. During this period, we have observed interesting diurnal rhythmic traits of *N. attenuata* and time-of-day dependent ecological interactions. For example, *N. attenuata* interacts with different groups of herbivores which are either day-active (such as grasshoppers, mirids and *Manduca* larvae) or night-active (such as noctuid larvae and tree crickets) and produces different chemicals that function as direct defenses against these herbivores or function as indirect defenses and attract of predators of the herbivores [49]. Recently, we showed that tissue specific diurnal rhythm of metabolites and its related transcripts in *N. attenuata* changes in response to herbivore attack of a specialist, *M. sexta* larvae [35]. More than 15% of total metabolites that we measured in leaf and root shows diurnal patterns and some of them have been demonstrated to function as plant defenses against herbivore attack. Goodspeed *et al.* [50] have recently reported that feeding behavior of *Tricoplusia ni* is predicted by the circadian clock in its host plant *Arabidopsis* and it increases anti-herbivore defense of *Arabidopsis*. All of these interactions provide a rich arena in which to explore the molecular mechanism of how the circadian clock regulates plant-insect interactions.

Conclusions

We identified three core circadian clock components in *N. attenuata* based on the gene expression and phenotypic alterations in lines silenced or overexpressed in the components. This work provides the foundation for the manipulation the ecological roles of the circadian clock in nature. As jet travel has revealed the depth of circadian-regulated processes in humans, circadian mutants of *N. attenuata* will be used to unravel the ecological functions of the clock in plant-environment, plant-plant, and plant-insect interactions in nature.

Methods

Plant growth condition

For all experiments we used *Nicotiana attenuata* Torr. Ex. Wats (Solanaceae) plants (31st inbred generation), wild-type (WT) originating from a population in Utah. Seeds were sterilized and germinated on Petri Dishes with Gamborg's B5 media as described in [43]. Petri dishes with 20 seeds were kept under long day conditions (LD, 16 h light/ 8 h dark).

To examine the free running period, one group of seedlings 15 days after germination was moved into constant light conditions for the 3 days of sampling and the other group of seedlings was grown under LD. Three biological replicates (10 seedlings pooled per replicate) were harvested every 2 h for 3 days and immediately frozen in liquid nitrogen.

RNA isolation and gene expression

Total RNA was extracted using TRIZOL reagent and cDNA was synthesized from 500 ng of total RNA using RT-minus kit (Fermentas, Burlington, Canada). The quantitative RT-PCR analyses were performed on a Stratagene MX3005P (Agilent Technologies, Santa Clara, CA, USA) employing SYBR Green kits (Eurogentec, Cologne, Germany). Primers were designed based on sequences from *Arabidopsis thaliana* retrieved from TAIR website and a *N. attenuata* transcript library (NCBI GEO Database accession number GSE30287). The sequences of qRT-PCR primer pairs for *NaLHY, NaTOC1, NaZTL* and *NaFKF1* are listed in Additional file 1. Transcript abundance expressed relative to the expression of *N. attenuata ELONGATION FACTOR* gene.

Phylogenetic analysis

The identity of the circadian genes *NaLHY* (NCBI accession number JQ424913), *NaTOC1* (Accession number JQ424914), *NaZTL* (Accession number JQ424912) and *NaFKF1/ADO3* (Accession number JQ424915) was determined by sequencing. A standard PCR was performed to obtain the full length cDNA and the list of primers used for this analysis is in Additional file 1. PCR products were subcloned for amplification using a CloneJET PCR Cloning kit (Fermentas).

The amino acid sequences of the circadian genes *NaLHY, NaTOC1, NaZTL* and *NaFKF1/ADO3* were deduced from cDNA sequences and aligned using the Geneious program V5.3 (http://www.geneious.com). The numbers of amino acid substitutions were estimated by a Jukes-Cantor model using a BLOSUM 62 matrix, through a global alignment with free end gaps option. A phylogenetic tree was reconstructed by the Unweighted Pair Group Method with Arithmetic Mean (UPGMA) method. These analyses were performed using Geneious software V5.3.

Overexpression of *NaLHY* and *NaZTL* in *arabidopsis*

All *A. thaliana* lines used were in the Col-0 background. Plants were grown in a controlled culture room at 22°C with a relative humidity of 55% under long-day (LD) conditions (16-h light/8-h dark) with white light illumination (120 µmol photons/m^{2}s) provided by fluorescent FLR40D/A tubes (Osram). To produce transgenic plants overexpressing the *NaLHY* and *NaZTL* genes, full-length cDNAs were subcloned into the binary pB2GW7 vector under the control of the CaMV 35S promoter (Invitrogen, Carlsbad, CA, USA). *Agrobacterium tumefaciens*–mediated *Arabidopsis* transformation was performed according to a modified floral dip method (Clough and Bent, 1998). T_2 transgenic plants harboring a single T-DNA insertion were used in subsequent assays including hypocotyl length and flowering time measurements. Transformation with *NaTOC1* was carried out but for unknown technical reasons we were not possible to regenerate plants from it. Selected lines were checked by RT-PCR for the *NaLHY* and *NaZTL* overexpression (Additional file 5), using *A. thaliana* tubulin (TUB) as a reference gene.

Yeast-two-hybrid

Yeast two-hybrid assays were performed using the Matchmaker™ system (Clontech, Palo Alto, CA, USA). Primers used for amplifying *NaTOC1*, *AtTOC1*, *NaZTL*, and *AtZTL* were described in Additional file 1. Each RT-PCR product was digested with restriction enzymes (*Eco*RI, *Xma*I for NaTOC1 and AtTOC1, *Nco*I, *Xma*I for NaZTL, *Nde*I, *Bam*HI for AtZTL). Digested full-length transcripts of *TOC1*s and *ZTL*s were subcloned into pGADT7 and pGBKT7, respectively. The yeast strain AH109 (leucine-, tryptophan-, histidine-, adenine-) contained *lacZ* reporter gene was co-transformed with the indicated vectors in Figure 5A. Transformation was conducted according to the manufacturer's instructions (Clontech). Single colonies obtained on growth medium lacking Leu, Trp were inoculated on a medium without Leu, Trp, His, Ade and used to measure β-galactosidase activity described in the instructions (Clontech).

Silencing of *NaTOC1* in *N. attenuata*

A sequence fragment of *NaTOC1* cDNA was inserted into the pSOL8 transformation vector as an inverted repeat construct driven by the CaMV 35S promoter [43,44]. The *NaTOC1* vector was transformed into *N. attenuata* WT plants using *Agrobacterium tumefaciens* mediated transformation; ploidy was determined on T_0 plants as described by Gase et al. [44], allowing for the selection of only diploid transformed lines. Homozygosity was confirmed on T_2 plants by hygromycin resistance screening after which 6 transformed lines were selected and transferred to the glasshouse for further growth at conditions described in Krügel et al. [43]. Gene expression levels of *NaTOC1* were determined by qPCR from tissue of selected T_2 plants and wild-type plants collected at ZT 12 h.

Authors' contributions

FY, PS, SK designed experiments and carried out the lab work, JR transformed *35S:NaLHY* and *35S:NaZTL* construct into *Arabidopsis*, CP, ITB and SK conceived the project and oversaw the research. All authors wrote, read and approved the final manuscript.

Acknowledgements

We thank the Max Planck Society for financial support. This work was also supported by the Leaping Research Program (20110016440) provided by the National Research Foundation of Korea, the Next-Generation BioGreen 21 program (Plant Molecular Breeding Center No. PJ008103) provided by the Rural Development Administration, and by grants from the Plant Signaling Network Research Center (20110001099), the National Research Foundation of Korea (20110027355), and the Agricultural R & D Promotion Center (309017–03), Korea Ministry for Food, Agriculture, Forestry and Fisheries.

References

1. Dunlap JC: **Genetic and molecular analysis of circadian rhythms.**

   Annu Rev Genet 1996, **30**(1)**:**579-601.

2. Sanchez A, Shin J, Davis SJ: **Abiotic stress and the plant circadian clock.**

   Plant Signaling Behav. 2011, **6**(2)**:**223-231.

3. Yakir E, Hilman D, Harir Y, Green RM: **Regulation of output from the plant circadian clock.**

FEBS J 2007, **274**(2):335-345.

4. Barak S, Tobin EM, Green RM, Andronis C, Sugano S: **All in good time: the *Arabidopsis* circadian clock.**

   Trends Plant Sci 2000, **5**(12):517-522.

5. Golden SS, Canales SR: **Cyanobacterial circadian clocks - timing is everything.**

   Nat Rev Microbiol 2003, **1**(3):191-199.

6. Lombardi LM, Brody S: **Circadian rhythms in *neurospora crassa*: clock gene homologues in fungi.**

   Fungal Genet Biol 2005, **42**(11):887-892.

7. Panda S, Antoch MP, Miller BH, Su AI, Schook AB, Straume M, Schultz PG, Kay SA, Takahashi JS, Hogenesch JB: **Coordinated transcription of key pathways in the mouse by the circadian clock.**

   Cell 2002, **109**(3):307-320.

8. Albrecht U, Eichele G: **The mammalian circadian clock.**

   Curr Opin Genet Dev 2003, **13**(3):271-277.

9. Loros JJ, Dunlap JC: **Genetic and molecular analysis of circadian rhythms in *N. eurospora*.**

   Annu Rev Physiol 2001, **63**(1):757-794.

10. Williams JA, Sehgal A: **Molecular components of the circadian system in drosophila.**

    Annu Rev Physiol 2001, **63**(1):729-755.

11. Ueda HR: **Systems biology flowering in the plant clock field.**

    Mol Syst Biol 2006, **2**:60.

12. Wang Z-Y, Tobin EM: **Constitutive expression of the *CIRCADIAN CLOCK ASSOCIATED 1 (CCA1)* gene disrupts circadian rhythms and suppresses its own expression.**

    Cell 1998, **93**(7):1207-1217.

13. Gendron JM, Pruneda-Paz JL, Doherty CJ, Gross AM, Kang SE, Kay SA: *Arabidopsis* **circadian clock protein, TOC1, is a DNA-binding transcription factor.**

    Proc Natl Acad Sci USA 2012, **109**(8):3167-3172.

14. Pokhilko A, Fernández AP, Edwards KD, Southern MM, Halliday KJ, Millar AJ: **The clock gene circuit in *Arabidopsis* includes a repressilator with additional feedback loops.**

    Mol Syst Biol 2012, **8**:574.

15. Alabadí D, Oyama T, Yanovsky MJ, Harmon FG, Más P, Kay SA: **Reciprocal regulation between TOC1 and LHY/CCA1 within the *Arabidopsis* circadian clock.**

    Science 2001, **293**(5531):880-883.

16. Farré EM, Harmer SL, Harmon FG, Yanovsky MJ, Kay SA: **Overlapping and distinct roles of PRR7 and PRR9 in the *Arabidopsis* circadian clock.**

    Curr Biol 2005, **15**(1):47-54.

17. Salomé PA, McClung CR: ***PSEUDO-RESPONSE REGULATOR 7* and *9* are partially redundant genes essential for the temperature responsiveness of the *Arabidopsis* circadian clock.**

    Plant Cell 2005, **17**(3):791-803.

18. Mas P, Kim W-Y, Somers DE, Kay SA: **Targeted degradation of TOC1 by ZTL modulates circadian function in *Arabidopsis thaliana*.**

    Nature 2003, **426**(6966):567-570.

19. Imaizumi T, Schultz TF, Harmon FG, Ho LA, Kay SA: **FKF1 F-box protein mediates cyclic degradation of a repressor of CONSTANS in** *Arabidopsis*.

    Science 2005, **309**(5732):293-297.

20. Sawa M, Nusinow DA, Kay SA, Imaizumi T: **FKF1 and GIGANTEA complex formation is required for day-length measurement in** *Arabidopsis*.

    Science 2007, **318**(5848):261-265.

21. Baudry A, Ito S, Song YH, Strait AA, Kiba T, Lu S, Henriques R, Pruneda-Paz JL, Chua N-H, Tobin EM, *et al.*: **F-box proteins FKF1 and LKP2 act in concert with ZEITLUPE to control** *Arabidopsis* **clock progression.**

    Plant Cell 2010, **22**(3):606-622.

22. Wang X, Wu L, Zhang S, Ku L, Wei X, Xie L, Chen Y: **Robust expression and association of ZmCCA1 with circadian rhythms in maize.**

    Plant Cell Rep 2011, **30**(7):1261-1272.

23. Takata N, Saito S, Tanaka Saito C, Nanjo T, Shinohara K, Uemura M: **Molecular phylogeny and expression of poplar circadian clock genes,** *LHY1* **and** *LHY2*.

    New Phytol 2009, **181**(4):808-819.

24. Liu H, Wang H, Gao P, Xu J, Xu T, Wang J, Wang B, Lin C, Fu YF: **Analysis of clock gene homologs using unifoliolates as target organs in soybean (***Glycine max***).**

    J Plant Physiol 2009, **166**(3):278-289.

25. Murakami M, Tago Y, Yamashino T, Mizuno T: **Comparative overviews of clock-associated genes of** *arabidopsis thaliana* **and** *oryza sativa*.

    Plant Cell Physiol 2007, **48**(1):110-121.

26. Izawa T, Oikawa T, Sugiyama N, Tanisaka T, Yano M, Shimamoto K: **Phytochrome mediates the external light signal to repress FT orthologs in photoperiodic flowering of rice.**

Genes Dev 2002, **16**(15):2006-2020.

27. Kaldis A-D, Kousidis P, Kesanopoulos K, Prombona A: **Light and circadian regulation in the expression of *LHY* and *lhcb* genes in *phaseolus vulgaris*.**

    Plant Mol Biol 2003, **52**(5):981-997.

28. Miwa K, Serikawa M, Suzuki S, Kondo T, Oyama T: **Conserved expression profiles of circadian clock-related genes in two lemna species showing long-day and short-day photoperiodic flowering responses.**

    Plant Cell Physiol 2006, **47**(5):601-612.

29. Ramos A, Pérez-Solís E, Ibáñez C, Casado R, Collada C, Gómez L, Aragoncillo C, Allona I: **Winter disruption of the circadian clock in chestnut.**

    Proc Natl Acad Sci USA 2005, **102**(19):7037-7042.

30. Taylor A, Massiah AJ, Thomas B: **Conservation of *arabidopsis thaliana* photoperiodic flowering time genes in onion (*allium cepa* L.).**

    Plant Cell Physiol 2010, **51**(10):1638-1647.

31. Xue ZG, Zhang XM, Lei CF, Chen XJ, Fu YF: **Molecular cloning and functional analysis of one ZEITLUPE homolog GmZTL3 in soybean.**

    Mol Biol Rep 2011, **39**(2):1411-1418.

32. Dodd AN, Salathia N, Hall A, Kévei E, Tóth R, Nagy F, Hibberd JM, Millar AJ, Webb AAR: **Plant circadian clocks increase photosynthesis, growth, survival, and competitive advantage.**

    Science 2005, **309**(5734):630-633.

33. Yerushalmi S, Yakir E, Green RM: **Circadian clocks and adaptation in *Arabidopsis*.**

    Mol Ecol 2011, **20**(6):1155-1165.

34. Green RM, Tingay S, Wang ZY, Tobin EM: **Circadian rhythms confer a higher level of fitness to *Arabidopsis* plants.**

Plant Physiol 2002, **129**(2):576-584.

35. Kim SG, Yon F, Gaquerel E, Gulati J, Baldwin IT: **Tissue specific diurnal rhythms of metabolites and their regulation during herbivore attack in a native tobacco.**

    Nicotiana attenuata. PLoS One 2011, **6**(10):e26214.

36. Mizoguchi T, Wheatley K, Hanzawa Y, Wright L, Mizoguchi M, Song H-R, Carré IA, Coupland G: ***LHY* and *CCA1* are partially redundant genes required to maintain circadian rhythms in *Arabidopsis*.**

    Dev Cell 2002, **2**(5):629-641.

37. Schaffer R, Ramsay N, Samach A, Corden S, Putterill J, Carré IA, Coupland G: **The late elongated hypocotyl mutation of *arabidopsis* disrupts circadian rhythms and the photoperiodic control of flowering.**

    Cell 1998, **93**(7):1219-1229.

38. Matsushika A, Makino S, Kojima M, Mizuno T: **Circadian waves of expression of the APRR1/TOC1 family of pseudo-response regulators in *arabidopsis thaliana*: insight into the plant circadian clock.**

    Plant Cell Physiol 2000, **41**(9):1002-1012.

39. Somers DE, Webb A, Pearson M, Kay SA: **The short-period mutant, *toc1-1*, alters circadian clock regulation of multiple outputs throughout development in *arabidopsis thaliana*.**

    Development 1998, **125**(3):485-494.

40. Millar AJ, Carre IA, Strayer CA, Chua NH, Kay SA: **Circadian clock mutants in *arabidopsis* identified by luciferase imaging.**

    Science 1995, **267**(5201):1161.

41. Somers DE, Schultz TF, Milnamow M, Kay SA: **ZEITLUPE encodes a novel clock-associated PAS protein from *arabidopsis*.**

    Cell 2000, **101**(3):319-329.

42. Somers DE, Kim WY, Geng R: **The F-box protein ZEITLUPE confers dosage-dependent control on the circadian clock, photomorphogenesis, and flowering time.**

    Plant Cell 2004, **16**(3):769-782.

43. Krügel T, Lim M, Gase K, Halitschke R, Baldwin IT: **Agrobacterium - mediated transformation of Nicotiana attenuata, a model ecological expression system.**

    Chemoecology 2002, **12**(4):177-183.

44. Gase K, Weinhold A, Bozorov T, Schuck S, Baldwin IT: **Efficient screening of transgenic plant lines for ecological research.**

    Mol Ecol Resour 2011, **11**(5):890-902.

45. Niwa Y, Ito S, Nakamichi N, Mizoguchi T, Niinuma K, Yamashino T, Mizuno T: **Genetic linkages of the circadian clock-associated genes, *TOC1*, *CCA1* and *LHY*, in the photoperiodic control of flowering time in *Arabidopsis thaliana*.**

    Plant Cell Physiol 2007, **48**(7):925-937.

46. Xu X, Xie Q, McClung CR: **Robust circadian rhythms of gene expression in *brassica rapa* tissue culture.**

    Plant Physiol 2010, **153**(2):841-850.

47. McClung CR: **A modern circadian clock in the common angiosperm ancestor of monocots and eudicots.**

    BMC Biol 2010, **8**(1):55.

48. Michael TP, Salomé PA, Yu HJ, Spencer TR, Sharp EL, McPeek MA, Alonso JM, Ecker JR, McClung CR: **Enhanced fitness conferred by naturally occurring variation in the circadian clock.**

    Science 2003, **302**(5647):1049.

49. Wu J, Baldwin IT: **New insights into plant responses to the attack from insect herbivores.**

    Annu Rev Genet 2010, **44**:1-24.

50. Goodspeed D, Chehab EW, Min-Venditti A, Braam J, Covington MF: ***Arabidopsis* synchronizes jasmonate-mediated defense with insect circadian behavior.**

    Proc Natl Acad Sci U S A 2012, **109**(12):4674-4677.

Additional file 1: List of primers used for transcript profiling and full-length cloning of circadian clock genes in *N. attenuata* and *A. thaliana*.

Gene	Description	Forward Primer	Reverse Primer
LHY	qPCR	CACTCTTTTCAAGGAAGGTG	GTCGAAGGTGTTACAAGAGC
TOC1	qPCR	ATCGTAGAACGGCAGCACTT	TCACAAACTGTCCCCTCACA
ZTL	qPCR	CCCTATTGACTCGCTTCTGC	GCCAAGGACTTCTTCAGCAC
FKF1	qPCR	ACAAGCCTACATGGAGAGAA	CCTCCAAGTCAATCGTGTAT
LHY	sequencing	ATGGACCCTTATTCCTCTGG	TCAAATAGAAGCTTCTCCTTCC
TOC1	sequencing	ATGGAGAAGAGTGAGATTGTTAAG	TCATAGACGCATCGATGGATC
ZTL	sequencing	ATGGAGTGGGACAGTAACTCG	TTATTCATATGGCAAGCTCGC
FKF1	sequencing	ATGGAAGGAGGAGGAGGAAAG	CATCATGCATCAGAATCTTGCT
TOC1	Yeast two-hybrid	GGGAATTCATGGAGAAGAGTGAGATTGT	GGCCCGGGTCATAGACGCATCGATGGAT
ZTL	Yeast two-hybrid	GGCCATGGAGATGGAGTGGGACAGTAACTC	GGCCCGGGTTATTCATATGGCAAGCTCG
AtTOC1	Yeast two-hybrid	GGGAATTCATGGATTTGAACGGTGAGTG	GGCCCGGGTCAAGTTCCCAAAGCATCATC
AtZTL	Yeast two-hybrid	GGCATATGATGGAGTGGGACAGTGGTTC	GGGGATCCCTAATGAGGAAGAAAGAAGAAGAAG

Additional file 2: Protein alignments of circadian clock genes in *N. attenuata*, Arabidopsis and rice. Full-length amino acid sequences were aligned using the Geneious software.

(A) LHY

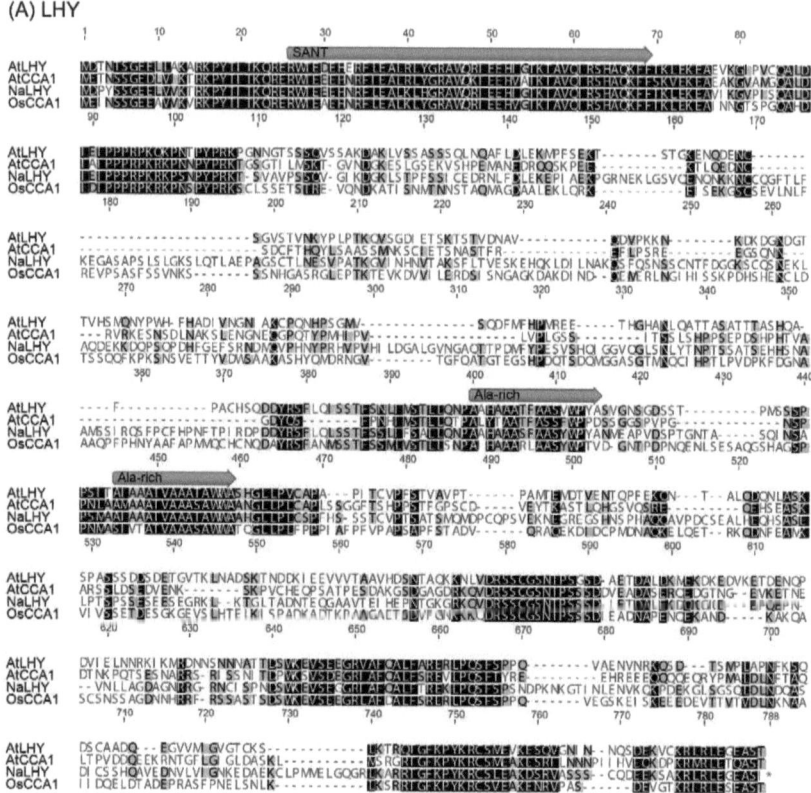

(B) TOC1

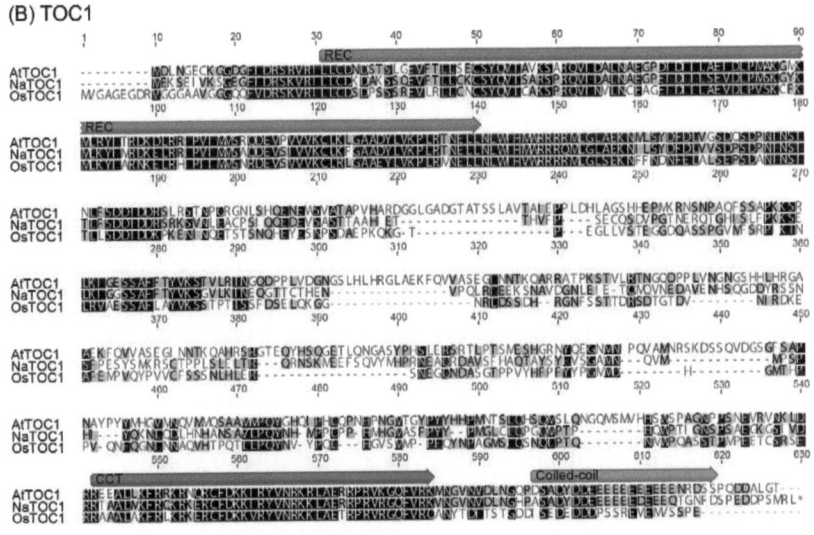

(C) ZTL

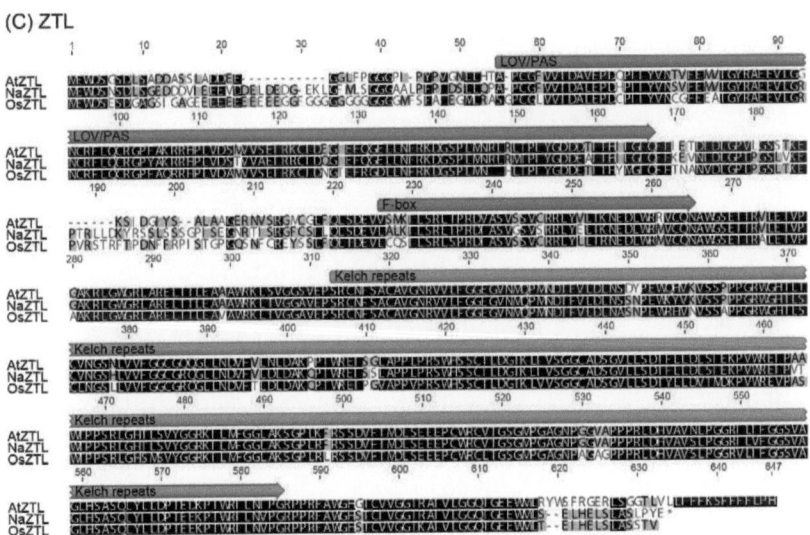

(D) FKF1/ADO3

Additional file 3: Phylogenetic trees of circadian clock genes in several plant species. Phylogenetic trees of (A) LHY/CCA1, (B) TOC1, (C) ZTL and (D) FKF1/ADO3. Full-length amino acid sequences were aligned using the Geneious software. Phylogenetic trees were constructed with available sequences from 3 major taxonomical groups: eudicots, monocots and one moss *Selaginella moellendorffii* for TOC1 and ZTL trees. Unweighted Pair Group Method with the Arithmetic Mean (UPGMA) method from the numbers of amino acid substitutions by applying the Jukes-Cantor model was used. The scale bar represents the number of amino acid substitutions per site. Ac, *Allium cepa*; At, *Arabidopsis thaliana*; Cs, *Castanea sativa*; Gm, *Glycine max*; In, *Ipomoea nil*; Mc, *Mesembryanthemum crystallinum*; Na, *Nicotiana attenuata*; Os, *Oryza sativa*; Pn, *Populus nigra*; Pt, *Populus trichocarpa*; Pv, *Phaseolus vulgaris*; Sb, *Sorghum bicolor*; Sl, *Solanum lycopersicum*; Sm, *Selaginella oellendorffii*; Vv, *Vitis vinifera*; Ta, *Triticum aestivum*; Zm, *Zea mays*.

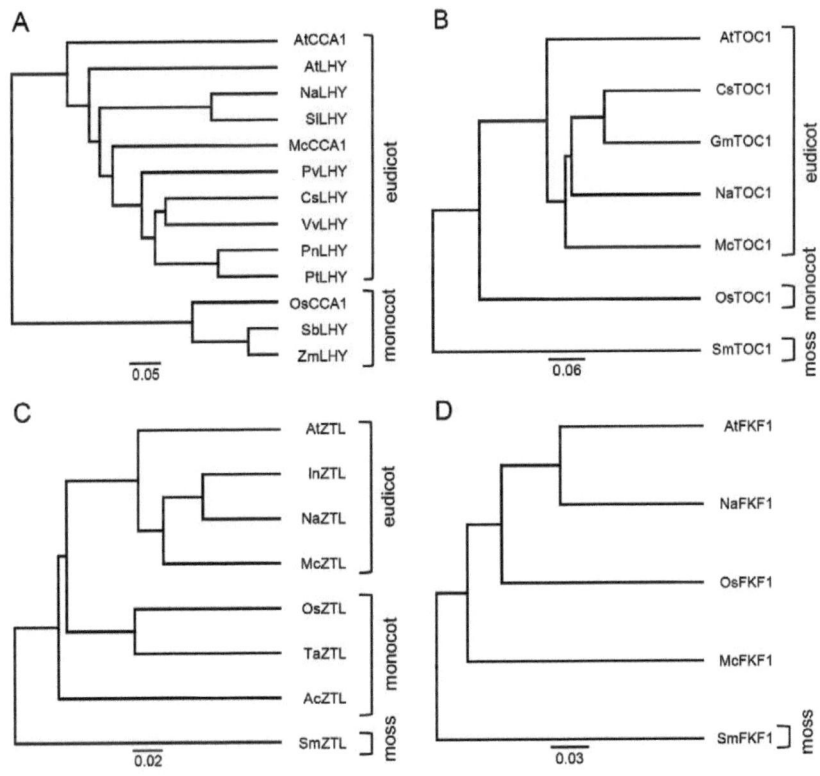

Additional file 4: Circadian rhythm of theNaFKF1/ADO3expression in N. attenuata. (A) Mean (± SE) levels of transcript abundance of NaFKF1/ADO3were examined using our time course microarray data (accession number GSE30287). Wild-type *N. attenuata* plants were harvested every 4 h for one day from leaves (n = 3) and roots (n = 3). After RNA isolation, each sample was hybridized on Agilent single color technology arrays designed from the *N. attenuata* transcriptome (accession number GPL13527). (B) Mean (± SE) levels of relative transcript abundance ofNaFKF1/ADO3in wild-type N. attenuataseedling (n = 20) at each harvest time for two days under long day condition (LD, gray lines, 16 h light/ 8 h dark) and continuous light condition (LL, black line). Transcript levels were determined by qPCR and *N. attenuata* Elongation Factor was used as reference gene. Gray boxes depict the dark period of LD.

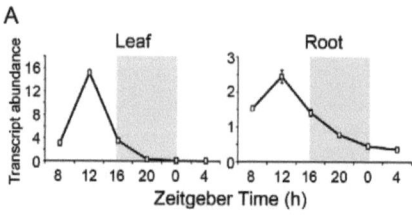

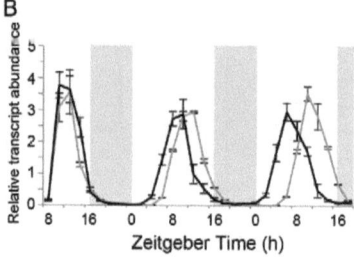

Additional file 5: Overexpression of NaLHY and NaZTL transcripts in Arabidopsis transformants. Transcript levels were measured by RT-PCR. Primers (see Additional file 1) were designed for the specific detection of NaLHY or NaZTL transcripts, not for Arabidopsis LHY or ZTL. Plants grown under LD were harvested at ZT 8 h. A tubulin gene (TUB) was included as a control.

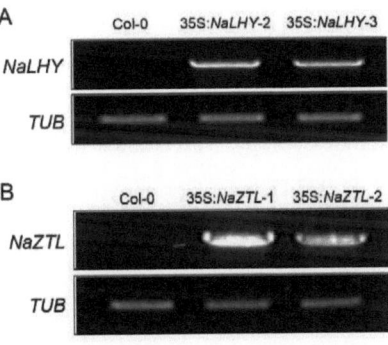

Chapter 4 – Manuscript II

Nicotiana attenuata **LHY and ZTL regulate circadian rhythms in flowers**

Felipe Yon, Youngsung Joo, Eva Rothe, Ian T. Baldwin, Sang-Gyu Kim[1]

Department of Molecular Ecology, Max Planck Institute for Chemical Ecology, Hans-Knöll-Straße 8, D-07745 Jena, Germany.

Abstract

The rhythmic opening/closing movement and volatile emissions of flowers attract pollinators at specific times. That these rhythms are maintained under constant light or dark conditions suggests a circadian clock is involved. A forward and reverse genetic approach allows for the identification of core circadian clock components in *Arabidopsis thaliana*. However, the role of these core clock components for floral rhythms has remained untested in Arabidopsis flowers, likely due to their weak diurnal rhythms. The wild tobacco *Nicotiana attenuata* flowers open at night, emit benzyl acetone scents, and move vertically through a 140° arc when engaged in opportunistic outcrossing. To examine the role of the clock components for floral rhythms, we generated transgenic *N. attenuata* lines silenced in the homologous genes of Arabidopsis *LATE ELONGATED HYPOCOTYL* (*LHY*) and *ZEITLUPE* (*ZTL*). The silencing of

NaLHY and *NaZTL* altered the transcript accumulation of *CHLOROPHYLL A/B BINDING PROTEINS 2* in seedlings exposed to constant light conditions and strongly affected floral rhythms under long day conditions, demonstrating that conserved clock components coordinate these floral rhythms.

Introduction

Linnaeus (1751) designed a garden known as the "flower clock," comprised of different plant species with unique flower opening and closing times. The opening of dandelion (*Taraxacum officinale*) flowers in his garden indicated morning, while the opening of *Mirabilis dichotoma* flowers meant it was around 4 o'clock in the afternoon. Many flowering plants also emit floral scents at specific times during a day. *Cestrum nocturnum* (night-blooming jasmine) (Overland 1960), *Nicotiana sylvestris* and *Nicotiana suaveolens* (Kolosova *et al.* 2001, Loughrin *et al.* 1991) emit a bouquet of floral scents at night, and *Antirrhinum majus* (snapdragon) flowers emit methyl benzoate in the afternoon (Kolosova, Gorenstein, Kish and Dudareva 2001). These famous examples show that flowering plants have characteristic rhythms which synchronize with environmental factors such as the active times of their pollinators' (Fründ *et al.* 2011, Somers 1999). In addition, classical experiments demonstrated the retention of floral rhythms under constant light (LL) or dark conditions, suggesting that an internal biological clock, called a circadian clock, regulates flower opening as well as the emission of floral volatiles (Bunning 1956, Kolosova, Gorenstein, Kish and Dudareva 2001, Loughrin, Hamilton-Kemp, Andersen and Hildebrand 1991, Overland 1960, Sweeney 1963, van Doorn and Van Meeteren 2003).

The plant circadian clock has been intensively elucidated in the genetic model species *Arabidopsis thaliana* (Nagel and Kay 2012). Forward and reverse genetic approaches have revealed that this clock consists of transcriptional and post-translational feedback loops. In Arabidopsis, two morning-expressed MYB transcription factors, LATE ELONGATED HYPOCOTYL (LHY) and CIRCADIAN CLOCK ASSOCIATED 1 (CCA1), bind to the promoter of the evening component, TIMING OF CAB EXPRESSION 1 (TOC1, also called PSEUDO-RESPONSE REGULATOR 1, PRR1) to repress *TOC1* transcription during the day. Near dusk, the positive regulator, REVEILLE8, induces the expression of *TOC1* transcript (Hsu *et al.* 2013), and TOC1 protein suppresses the expression of *LHY* and *CCA1* transcripts, establishing a transcriptional negative feedback loop (Gendron *et al.* 2012, Huang *et al.* 2012). Post-translational regulation also fine-tunes the plant circadian clock. ZEITLUPE (ZTL) protein physically binds to TOC1 and PRR5 proteins under dark conditions, resulting in the degradation of TOC1 and PRR5 proteins (Kiba *et al.* 2007, Kim *et al.* 2007, Mas *et al.* 2003). Two homologous proteins of ZTL, FLAVIN BINDING KELCH REPEAT F-BOX 1 and LOV KELCH REPEAT PROTEIN 2, also adjust the protein stability of the clock components (Baudry *et al.* 2010, Nelson *et al.* 2000, Sawa *et al.* 2007). Like ZTL, the FKF1 protein interacts with GIGANTEA in a blue-light-dependent manner and connects the clock to photoperiodic flowering of Arabidopsis (Sawa, Nusinow, Kay and Imaizumi 2007).

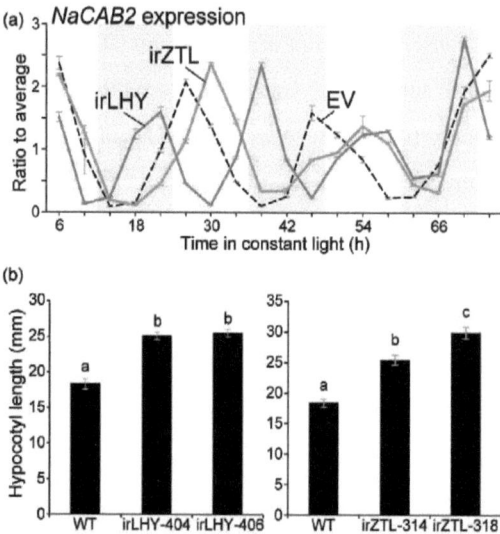

Figure 1. Silencing *NaLHY* and *NaZTL* alters an internal rhythm in seedlings. (A) Transcript accumulation of *CAB2* in seedlings of EV, irLHY, and irZTL grown under 12 h light and 12 h dark conditions and then exposed to constant light (LL) conditions. Seedlings were harvested every 4 h for three days. The ratio to average was calculated by dividing the average transcript levels at the each time point by the average levels of the same transcript across all time points. Gray boxes indicate the subjective dark period of LL conditions. (B) Mean (±SE) length of hypocotyl in EV, irLHY, and irZTL seedlings grown under the dim light conditions. LHY, LATE ELONGATED HYPOCOTYL; ZTL, ZEITLUPE; EV, plant transformed with the empty-vector used to generate transgenic lines; irLHY, *NaLHY*-silenced line; irZTL, *NaZTL*-silenced line; CAB2, CHLOROPHYLL A/B BINDING PROTEINS 2

Altering the expression of these circadian clock genes has produced arrhythmic or disrhythmic plants; these plants show many defects in development (Nagel and Kay 2012) and defense (Goodspeed *et al.* 2012, Wang *et al.* 2011a). For instance, several daily rhythmic traits, such as stomata aperture, leaf movement and the expression of photosynthetic machinery are altered in clock-altered lines (Yakir *et al.* 2007). In addition, hypocotyl elongation, flowering time and biotic/abiotic defense are also regulated by the circadian clock and have been examined using clock-altered lines (Nagel and

Kay 2012, Seo *et al.* 2012, Wang, Barnaby, Tada, Li, Tor, Caldelari, Lee, Fu and Dong 2011a). However, to our knowledge, no one has demonstrated that the internal clock in flowers is the circadian clock whose molecular details are now known. Are the floral rhythms regulated by the known circadian clock components? This question is frequently noted in literatures as one that festers and remains to be rigorously tested (Nitta *et al.* 2010, van Doorn and Van Meeteren 2003, Yakir, Hilman, Harir and Green 2007).

To examine the influence of the core clock components on floral rhythms, we used the wild tobacco *Nicotiana attenuata*, which shows strong diurnal rhythms in flowers and whose plant-pollinator interactions have been well studied. *N. attenuata* produces self-compatible flowers which are visited by nocturnal hawkmoths (e.g. *Manduca sexta*) and day-active pollinators such as hummingbirds (Kessler *et al.* 2010). Approximately 95% of *N. attenuata* flowers open at night; at this time they emit a bouquet of volatiles, mainly benzyl acetone (BA), which attracts nocturnal hawkmoths (Kessler, Diezel and Baldwin 2010). These floral rhythms are repeated for two or three days, and then, if pollination occurs, corollas senesce, capsules develop and seeds mature.

In a previous study, we identified the *N. attenuata* LHY (NaLHY), NaTOC1, and NaZTL, which are the homologous proteins of Arabidopsis LHY, TOC1, and ZTL, respectively (Yon *et al.* 2012). The oscillating patterns of these genes under light-dark cycles and LL conditions are similar to those of Arabidopsis clock components. To examine the functional conservation of the Arabidopsis clock components, we generated the overexpression lines of *NaLHY* and *NaZTL* transcripts in Arabidopsis. These lines displayed elongated hypocotyls and late flowering compared to wild-type (WT) plants: phenotypes similar to those of Arabidopsis LHY- and ZTL-overexpressing lines (Schaffer *et al.* 1998, Somers *et al.* 2004). In addition, TOC1-ZTL interactions in Arabidopsis are also conserved in *N. attenuata*; the NaTOC1 protein binds

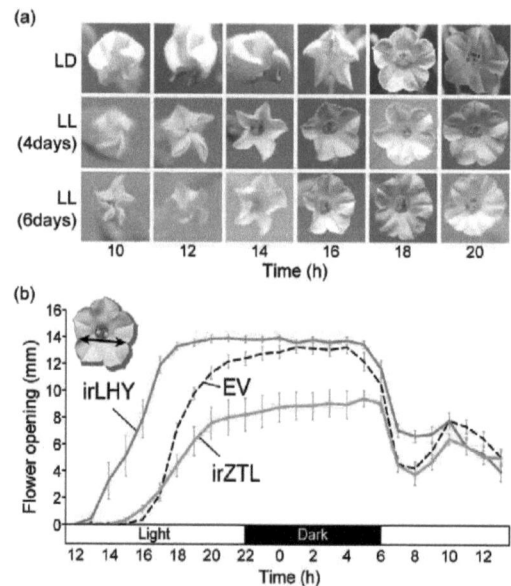

Figure 2. *N. attenuata* LHY and ZTL regulate flower opening. (A) Flower opening in wild-type *N. attenuata* grown under LD conditions and exposed to LL conditions for 4 days and 6 days. (B) Mean (±SE) distance between petal junctions on corolla limbs of EV, irLHY, and irZTL plants grown under LD conditions. LD, 16 h light and 8 h dark; LL, constant light.

NaZTL and Arabidopsis ZTL proteins as well, indicating that NaLHY, NaTOC1, and NaZTL are functionally homologous proteins of Arabidopsis (Yon, Seo, Ryu, Park, Baldwin and Kim 2012). In this study, we show that silencing *NaLHY* and *NaZTL* alters the internal rhythm of *N. attenuata* and three main diurnal rhythms of the flowers. In addition, transcript levels of *NaLHY* differ between corollas and pedicels.

Results

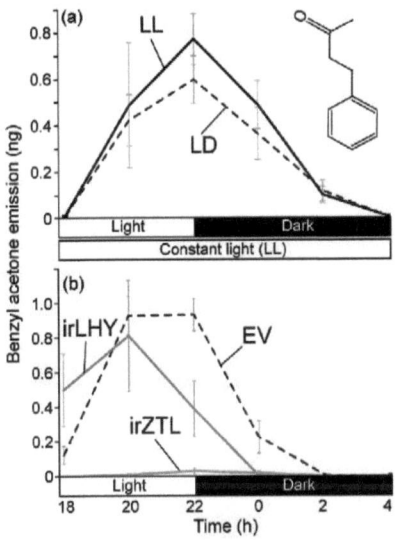

Figure 3. *N. attenuata* LHY and ZTL regulate attractive floral volatile, benzyl acetone (BA) emission from flowers. (A) Mean (±SE) levels of BA emission from wild-type plants under LD and LL conditions. We exposed LD-grown flowering plants to LL condition for 24 h and measured BA emission using a z-Nose™ instrument for real time measurements. (B) Mean (±SE) levels of BA emission from flowers in EV, irLHY, and irZTL plants grown under LD conditions. LD, 16 h light and 8 h dark; LL, constant light.

Silencing *NaLHY* and *NaZTL* alters the internal rhythm in *N. attenuata*.

We silenced the transcript levels of *NaLHY* and *NaZTL* in *N. attenuata* by transforming plants with gene-specific inverted-repeat (ir) constructs and identified more than two independent lines, each of which harbored a single insertion of the transformation construct and displayed more than 90% silencing efficiency at the peak expression times of the targeted gene (Figure S1). Empty vector-containing (EV) plants were used as controls to control for possible transformation effects, which were not observed. To examine the internal rhythm of these lines, we measured the transcript levels of *N. attenuata CHLOROPHYLL A/B BINDING PROTEINS 2* (*CAB2*) (Figure S2), which has

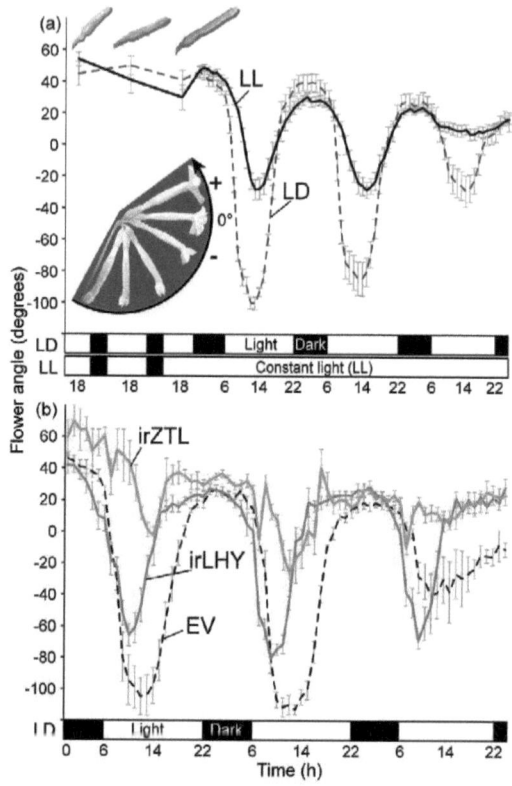

Figure 4. *N. attenuata* LHY and ZTL regulate vertical movement in flowers. (A) Mean (±SE) angles of flowers in *N. attenuata* wild-type plants under LD and LL conditions. Flower movement is initiated in the morning of the first-opening day and repeated over 2-3 days. Flower photos in (A) were taken at 6 different times in a day and merged after removing background colors using Adobe Photoshop. (B) Mean (±SE) angle of flowers in EV, irLHY, and irZTL plants grown under LD conditions. LD, 16 h light and 8 h dark; LL, constant light.

been frequently used to determine internal rhythms of the Arabidopsis clock-altered lines (Nagel and Kay 2012, Somers *et al.* 1998). Seedlings were grown under 12 h light and 12 h dark conditions for 12 days and exposed to LL conditions. After that, we collected samples of EV, irLHY (*NaLHY*-silenced line), and irZTL (*NaZTL*-silenced line) plants every 4 h for 3 days under LL conditions. As shown in Arabidopsis (Mizoguchi et al. 2002, Somers et al. 2000), the period of transcript oscillation of *NaCAB2* under LL conditions was shorten by silencing *NaLHY* compared to the period of *NaCAB2* oscillation in EV plants, and silencing *NaZTL* lengthened the period of *NaCAB2* oscillation (Figure 1a).

In a previous study, we showed that the overexpression of *NaLHY* and *NaZTL* in Arabidopsis seedlings results in an elongated hypocotyl compared to the hypocotyl in WT seedlings (Yon, Seo, Ryu, Park, Baldwin and Kim 2012). To test whether silencing *NaLHY* and *NaZTL* alters hypocotyl length in *N. attenuata*, we germinated the seeds under dim light conditions, and 10 days later we measured hypocotyl lengths of the lines. Seedlings of irLHY and irZTL displayed significantly increased hypocotyl lengths compared to seedlings of EV plants (Figure 1b). Taken together, these results indicate that silencing *NaLHY* and *NaZTL* alters internal rhythms of plants and suggest functional similarity of LHY and ZTL between *Arabidopsis* and *N. attenuata*.

NaLHY and NaZTL regulate flower opening.

To examine whether NaLHY and NaZTL regulate diurnal rhythms in flowers, we first examined flower-opening times in the lines. Flowers in *N. attenuata* started to open around 4 pm (Zietgeber Time ZT10) in plants under long day (LD, 16 h light and 8 h dark) conditions, and fully opened by 8 pm (ZT14) before dusk (Figure 2). Fully opened flowers displayed white flattened corolla limbs during the night (Figure 2) (Kessler, Diezel and Baldwin 2010). The flowers rapidly closed within 1 h of dawn the next day, but stopped in a half-opened position, which they retained over the day (Figure 2b). Under LL conditions, the timing of flower opening in *N. attenuata* remained unchanged (allowing for a 2 h difference) at least for 6 days (Figure 2a), as shown in several flowering plants (Bunning 1956, Overland 1960, Sweeney 1963, van Doorn and Van Meeteren 2003, Yakir, Hilman, Harir and Green 2007). Flowers from plants exposed to LL did not close well (Figure S3), suggesting that an internal clock in *N. attenuata* mainly regulates floral opening but not closing.

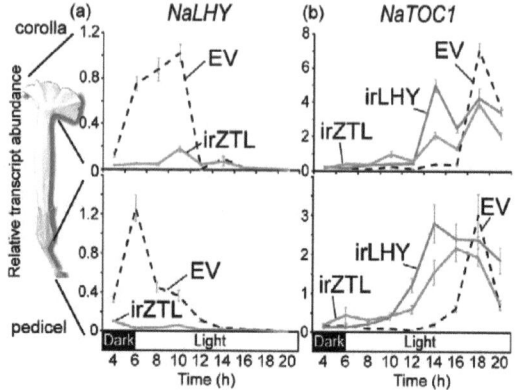

Figure 5. *NaLHY* and *NaTOC1* transcript expression in corolla limbs and pedicels of EV, irLHY, and irZTL plants. (A) Mean (± SE) levels of transcript accumulation of *NaLHY* in corolla limbs and pedicels of EV and irZTL. (B) Mean (± SE) levels of transcript accumulation of *NaTOC1* in corollas and pedicels of EV, irLHY, and irZTL.

Next, we analyzed the timing of flower opening and closing in irLHY and irZTL lines grown under LD conditions. The distance between the junctions on a corolla limb was measured to characterize the opening and closing of flowers (Figure 2b, inset). Flowers in irLHY lines began opening 2 h earlier than did EV flowers and reached full opening 2 h earlier at 6 pm (ZT12) (Figure 2b). Interestingly, irZTL flowers began opening at the same time as EV flowers but did not open completely: they were approximately 60% open compared to EV flowers, which were fully open (Figure 2b). By the next morning, irLHY and irZTL flowers closed rapidly within 1 h as did EV flowers, but the closing patterns of these flowers differed among the lines (Figure 2b).

NaLHY and NaZTL regulate floral scent emission.

N. attenuata flowers emit several volatiles to attract pollinators at night (Kessler and Baldwin 2007). The most abundant attractant, BA, is released from fully opened flowers: its release begins near dusk and lasts until the middle of the night (Figure 3) (Kessler, Diezel and Baldwin 2010). This diurnal emission is repeated for 2-3 days (Bhattacharya and Baldwin 2012), synchronized with flowers' opening/closing times. We first monitored BA emission every 2 h in the headspace of WT flowers under LD and LL conditions using a z-Nose^{TM} instrument for real-time measurements. For LL experiments, we exposed LD-grown plants to LL conditions 24 h before flowers opened and then measured BA amount. The pattern of BA emission from flowers under LL conditions was similar to the pattern of BA emission from flowers under LD conditions (Figure 3a), suggesting that BA emission in *N. attenuata* is regulated by an internal clock.

To determine whether NaLHY and NaZTL regulate the emission of floral volatiles, we monitored BA emission from the *NaLHY-* and *NaZTL*-silenced lines. BA emission from irLHY flowers started earlier but also declined earlier than did BA emission from EV flowers (Figure 3b). BA emission was correlated with early opening phenotypes, suggesting that rhythms in irLHY flowers shift to earlier times than do rhythms in EV flowers. Interestingly, BA emission was not detected from irZTL flowers (Figure 3b).

NaLHY and NaZTL regulate diurnal vertical movement of flowers.

N. attenuata flowers have an additional rhythmic trait. WT flowers in *N. attenuata* maintain an upright position approximately 40° from the horizontal axis before opening (Figure 4a). In the morning of the first opening day, flowers move to face down at more than 90° below the horizontal axis (Figure 4a). These flowers return to the upright position just before dusk (Figure 4a, inset), when they fully open and emit BA. By the next morning, flowers face down again and have closed their corollas. This vertical movement of flowers is repeated for 2-3 days under LD conditions, with diminishing movement of flowers in the third day.

To examine whether this diurnal movement is independent of light-dark cycles, we exposed LD-grown flowering plants to constant light (LL) conditions 24 h before flowers opened and measured the angle of flowers for three days (Figure 4a). Flowers exposed to LL conditions started to move downward when LD-grown flowers did, but the amplitude of movement in LL-exposed flowers (1st day, 39.2°; 2nd day, 28.1°; 3rd day, 6.0°) was reduced in comparison to that of flowers grown under LD conditions (1st day, 73.8°; 2nd day, 65.5°; 3rd day, 28.0°). The maximum upward angle in LL-exposed flowers was similar to the maximum angle in flowers grown under LD conditions. This result suggests that an endogenous clock regulates flower movement in *N. attenuata*, but that light signals are also needed to adjust the amplitude of the movement.

To clarify whether core clock components control this movement, we measured the angle of flowers in EV, irLHY, and irZTL. Silencing *LHY* and *ZTL* strongly altered flower movements in different ways (Figure 4b). The timing of the downward movement in irLHY lines during the first day was similar to the timing of the same movement in EV flowers, but irLHY flowers moved upward approximately 2 h earlier than did EV flowers, resulting in a smaller amplitude of flower movement in irLHY flowers (Figure 4b). This earlier vertical movement was correlated with flower opening and initial scent emission occurring 2 h earlier in irLHY plants than in EV plants. The period of movement in irLHY flowers (22.4h ± 0.2h) for the first two days was significantly shorter than that in EV flowers (23.5h ± 0.1h, $P < 0.05$, one-way ANOVA followed by Bonferroni *post hoc* tests). An alteration of the movement was also observed in irZTL lines; downward movement was almost abolished, but plants retained the weak diurnal pattern for the first two days (Figure 4b).

Expression of *NaLHY* and *NaTOC1* in flower tissues.

To test whether silencing *NaLHY* or *NaZTL* alters the clock gene expression in flowers, we collected corolla limbs, where flower opening and BA emission occurs, as well as pedicels which are primarily engaged in vertical movement of flowers. We measured the transcript levels of *NaLHY* in EV and irZTL plants and the transcript levels of *NaTOC1* in EV, irLHY, and irZTL plants every 2 h from 4 am (ZT22) to 8 pm (ZT14) under LD conditions. 12 flowers among 30 plants per lines at each time point were collected. The results revealed that the levels of *NaLHY* transcripts were reduced both in corolla limbs and pedicels of irZTL (Figure 5a). Silencing *NaLHY* and *NaZTL* induced the transcript expression of *NaTOC1* earlier than in EV both in corolla limbs and pedicels (Figure 5b). Interestingly, the transcript levels of *NaLHY* in EV plants differed between corolla limbs and pedicels. The transcript levels of *NaLHY* in pedicels (Figure 5) peaked at dawn and returned to the basal level within 4 h under LD conditions, which was similar to the *NaLHY* expression in leaves and roots (Yon, Seo, Ryu, Park, Baldwin and Kim 2012). However, *NaLHY* expression in corolla limbs did not return to the basal level for up to 6 h.

Discussion

Following the first scientific report on the effect of daily leaf movement in mimosa that were kept under constant dark conditions in 1729, several daily rhythms in plants have been examined (McClung 2006). Diurnal rhythms in flowers are one of the most popular examples known to both chronologists as well as non-scientists. Many reports including time-lapse movies and nature documentaries demonstrate that internal clocks regulate floral rhythms. However, these interesting floral rhythms have not been re-examined after the core circadian clock components were molecularly identified. Perhaps floral traits were not examined in clock-altered lines because researchers assumed that any trait exhibiting a diurnal rhythm under free-running conditions would be regulated by the clock system defined mainly in leaves. However, this assumption is not justified, considering the recent literatures on circadian rhythms. For example, the circadian clock system in roots of Arabidopsis is known to be different from the circadian clock system in the leaves, indicating the existence of the tissue-specific clock system in plants (James *et al.* 2008). In addition, redox cycles of peroxiredoxins mediate circadian rhythms in erythrocytes that lack transcription (Edgar *et al.* 2012, O'Neill and Reddy 2011), indicating that the clock gene expression is not required for all circadian rhythms. Here we revisit a set of floral traits previously thought to be under circadian control with knock-down lines of verified clock components and provide the first test of the hypothesis that 'diurnal rhythms in flowers are regulated by the circadian clock components' at a molecular level.

Homologous genes of the clock components elucidated in Arabidopsis have been identified in many eudicotyledonous and monocotyledonous plants (McClung 2013, Staiger *et al.* 2013): poplar (Filichkin *et al.* 2011, Takata *et al.* 2009), chestnut (Ramos *et al.* 2005), soybean (Liu *et al.* 2009, Xue *et al.* 2011), *Brassica rapa* (Lou *et al.* 2012, Xu *et al.* 2010), rice (Filichkin, Breton, Priest, Dharmawardhana, Jaiswal, Fox, Michael, Chory, Kay and Mockler 2011, Murakami *et al.* 2007), and maize (Khan *et al.* 2010, Wang *et al.* 2011b). The genome of green alga *Ostreococcus tauri* also contains TOC1- and CCA1-like genes (Corellou *et al.* 2009), suggesting that circadian clock genes are well conserved in plants. To extend the results of our previous study (Yon, Seo, Ryu, Park, Baldwin and Kim 2012), we provide new evidence that NaLHY and NaZTL are the functional homologues of Arabidopsis LHY and ZTL. Silencing *NaZTL* reduced the transcript accumulation of *NaLHY* under LD conditions (Figure 5), as shown in the Arabidopsis *ztl* mutants (Baudry, Ito, Song, Strait, Kiba, Lu, Henriques, Pruneda-Paz, Chua, Tobin, Kay and Imaizumi 2010, Somers, Kim and Geng 2004). The phase-shift of *NaTOC1* expression in irLHY

(Figure 5) is consistent with the phase-shift of *TOC1* expression in the Arabidopsis mutant *lhy-12* (suppressor mutation of *LHY*-overexpressed line) (Mizoguchi, Wheatley, Hanzawa, Wright, Mizoguchi, Song, Carré and Coupland 2002). In addition, *CAB2* expression in irLHY and irZTL seedlings was also similar to *CAB2* expression in Arabidopsis *lhy-12* and *ztl-3* plants.

Peak times of Arabidopsis *LHY* and *TOC1* transcript levels in leaves and roots (James, Monreal, Nimmo, Kelly, Herzyk, Jenkins and Nimmo 2008, Schaffer, Ramsay, Samach, Corden, Putterill, Carré and Coupland 1998, Strayer et al. 2000) are also well-conserved in *N. attenuata* leaves, roots, and flowers (Figure 5) (Yon, Seo, Ryu, Park, Baldwin and Kim 2012). The transcript levels of *NaLHY* and *NaTOC1* in corollas and pedicels peaked at dawn (6 h, ZT0) and at near dusk (18 h, ZT12), respectively, under LD conditions. However, the transcript levels of *NaLHY* in corollas declined more slowly than did these levels in pedicels (Figure 5) and leaves (Yon, Seo, Ryu, Park, Baldwin and Kim 2012), suggesting there is a flower-specific circadian clock in *N. attenuata*. In a previous study, we identified a homologous protein of Arabidopsis TOC1 and discovered a late-flowering phenotype of *NaTOC1*-silenced lines under LD conditions. Although the TOC1 protein is one of the major targets of ZTL, silencing NaTOC1 did not alter the floral rhythms under LD conditions (Figure S4), suggesting that the NaZTL protein regulates floral rhythms through other target proteins (e.g. PRR5 and GI) (Kiba, Henriques, Sakakibara and Chua 2007) or unknown flower-specific proteins.

The floral phenotypes of irZTL plants are not well explained by the long internal rhythms of irZTL, which was determined by *NaCAB2* expression under LL conditions (Figure 1). Incomplete opening, no BA emission, and small amplitude of vertical movement in irZTL flowers comprise a suite of arrhythmic traits. Internal rhythms of the clock-altered plants have been determined mainly by the expression patterns of *CAB2* or *COLD-CIRCADIAN*

RHYTHM-RNA BINDING 2 (*CCR2*) transcripts under constant light or dark conditions (Nagel and Kay 2012). For instance, the period of *CAB2* or *CCR2* transcripts in Arabidopsis *lhy-12* and TOC1 RNAi plants was shorter than their periods in WT plants under free-running conditions (Más *et al.* 2003, Mizoguchi, Wheatley, Hanzawa, Wright, Mizoguchi, Song, Carré and Coupland 2002, Strayer, Oyama, Schultz, Raman, Somers, Más, Panda, Kreps and Kay 2000), and the *ztl-1* mutation lengthens the period of *CAB2* and *CCR2* expression under LL conditions (Somers, Schultz, Milnamow and Kay 2000). However, the phenotypes of clock-altered plants are commonly found to be different or even opposite although these plants have same internal rhythms (Nagel and Kay 2012, Niwa *et al.* 2009), indicating that internal rhythms defined by the expression of a single reporter construct in a single tissue might not fully explain the complex traits regulated by the circadian clock.

Most insect-pollinated flowers have evolved special traits to attract pollinators (Raguso 2004), including the ability to synchronize flower rhythms with times when pollinators are active (van Doorn and Van Meeteren 2003). In nature, *N. attenuata* flowers produce morning- and night-opening flowers, which are synchronized with day-active or night-active pollinators, respectively (Kessler, Diezel and Baldwin 2010). The downward-facing movement of *N. attenuata* flowers likely prevents nectar from drying up during the day in its native habitats, in particular the Great Basin Desert, Utah, and the upward-facing movement might increase the accessibility of *M. sexta* moths during the night. In this study, we show that the conserved clock components regulate the circadian rhythms in flowers, which sustain the pollination services mediated by insects for many wild plants as well as in domesticated crops (Potts *et al.* 2010). We conclude that the circadian clock in flowers is the "battery" that makes the hands of Linnaeus's multi-species "flower clock" tick.

EXPERIMENTAL PROCEDURES

Plant growth conditions

We used *Nicotiana attenuata* Torr. Ex. Wats (Solanaceae) plants (30st inbred generation), which originated from a population in Utah. Seeds were sterilized and germinated on Petri dishes with Gamborg's B5 media as described in Krügel *et al.*(55). Petri dishes with 30 seeds were kept under long-day conditions (LD, 16h light and 8h dark) in a growth chamber (Percival, Perry, Iowa, USA) for 10 days, and seedlings were transferred to small pots (TEKU JP 3050 104 pots, Pöppelmann GmbH & Co. KG, Lohne, Germany) with Klasmann plug soil (Klasmann-Deilmann GmbH, Geesten, Germany) in the glasshouse. After 10 days, plants were transferred to 1L pots. The glasshouse growth conditions are described in Krügel *et al.* (55). For the light-dark and constant light treatment, two growth chambers (Microclima 1000, Snijders Scientific, Netherlands) were maintained at similar humidity and temperature conditions with the glasshouse conditions. To measure hypocotyl length, seedlings were grown on vertically oriented agar plates under dim light conditions for 10 days.

The silencing of *NaLHY*, *NaTOC1*, and *NaZTL* in *N. attenuata*

A specific fragment of *NaLHY* (NCBI accession number JQ424913), *NaTOC1* (JQ424914) and *NaZTL* (JQ424912) was independently inserted into the pSOL8 (for *NaTOC1* and *NaZTL*) and pRESC8 (for *NaLHY*) transformation vectors as an ir construct driven by the CaMV 35S promoter (56). These vectors were transformed into *N. attenuata* WT plants using *Agrobacterium tumefaciens*-mediated transformation, and diploid transformed lines were selected as described in Gase *et al.* (56). Homozygosity was confirmed in T_2 plants by hygromycin resistance, and selected lines were transferred to the glasshouse for further analysis. Transformed WT plants with an EV were used as controls for characterizing the transgenic lines. Gene expression levels of each silenced line were determined by qPCR from rosette leaf tissues of selected T_2 plants collected at ZT0 for irLHY, and at ZT12 for irZTL. Total RNA was extracted using the TRIzol reagent (Invitrogen) and 1 µg of total RNA of each sample was used to synthesize a single strand cDNA with reverse transcriptase (Fermentas). Quantitative real-time polymerase chain reaction (qPCR) was conducted with a Stratagene MX3005p instrument and SYBR Green kit (Eurogentec). The sequences of primers used for qPCR (NaLHY-F, CACTCTTTTCAAGGAAGGTG; NaLHY-R, GTCGAAGGTGTTACAAGAGC; NaTOC1-F, ATCGTAGAACGGCAGCACTT; NaTOC1-R, TCACAAACTGTCCCCTCACA; NaZTL-F, CCCTATTGACTCGCTTCTGC; NaZTL-R, GCCAAGGACTTCTTCAGCAC; NaCAB2-F, GCCGGAAAGGCAGTGAAAC; NaCAB2-R, ACCGGGTCTGCAAGATGATC) were designed by Geneious (Version 5.7.7, http://www.geneious.com). Finally, we selected two independent lines of the clock-silenced lines: irLHY404, irLHY-406, irTOC1-205, irTOC1-212, irZTL 314, and irZTL-318. All data shown in main figures were derived from irLHY-406 and irZTL-314 and the data from irLHY-404 and irZTL-318 were shown in Fig. S5.

Measurements of diurnal rhythms in flowers

Flower position was recorded at 1 h acquisition intervals using a time-lapse imaging setup, composed of a digital camera IXUS 400 (Canon, Tokyo, Japan) and its remote control software ZoomBrowser v5.6 (Canon, Tokyo, Japan). Selected flowers in photos were analyzed using the software Image Tools v.3.0 (UTHSCSA) and Tracker v.4.72 (Cabrillo College). Flower angles were measured with reference to the horizontal axis. Flower opening was measured using excised flowers with 6 to 12 biological replicates for each measurement. Photos were taken every 30 min using a time lapse imaging setup. To quantify the opening, the inner distance between opposite lobes was measured in pixels and converted to millimeters.

BA emission from flowers was measured using a portable gas chromatograph, z-Nose™ 4200 (Electronic Sensor Technology, Newbury Park, CA) in real time. To make a headspace trap, 50mL plastic tubes (Falcon Plastics) were cut in half, and the upper parts with a cap were used. A single hole was made in a cap to inject a needle into a headspace of flowers.

Analysis of flower rhythms and statistical test

Rhythmic parameters (period, phase, amplitude) were measured using the ARSER algorithm (57), which was designed to identify circadian rhythms in gene expression. To detect rhythms in expression data, ARSER first removes any linear trends from the data and then determines the period of the expression data. Finally, ARSER provides four rhythmic parameters: period, phase, amplitude, and mean, using harmonic regression analysis. To calculate amplitude of WT flower movement in plants under LD and LL conditions, flower angle data were divided into three parts (first, second, and third day) and time-series data in each part were concatenated before ARSER analysis. To measure the period and the amplitude of the clock-silenced lines, flower angle data from first and second day were used. After measuring period and amplitude of each flower by ARSER analysis, mean (±SE) values of each clock-silenced line were calculated.

All statistical tests were performed using R 2.12.1 (http://www.r-project.org/) and R-Studio (Version 0.96.316, http://www.rstudio.com/).

Acknowledgments

We thank Janet Grabengießer, and Lucas Cortés Llorca for technical assistance, Emily Wheeler for editing, Dr. Klaus Gase for designing ir-constructs, and Drs. Emmanuel Gaquerel, Meredith C. Schuman, Stefan Meldau, and Danny Kessler for critical comments on the manuscript. F. Yon, Y. Joo, S. Kim, and E. Rothe screened and characterized the transgenic lines, I.T. Baldwin and S. Kim designed the experiments and conceived of the project. All authors declare that they have no conflicts of interest. This work is supported by European Research Council advanced grant ClockworkGreen (No. 293926) to ITB, the Global Research Lab program (2012055546) from the National Research Foundation of Korea, and the Max Planck Society.

References

Baudry, A., Ito, S., Song, Y.H., Strait, A.A., Kiba, T., Lu, S., Henriques, R., Pruneda Paz, J.L., Chua, N.-H., Tobin, E.M., Kay, S.A. and Imaizumi, T. (2010) F-box proteins FKF1 and LKP2 act in concert with ZEITLUPE to control *Arabidopsis* clock progression. *Plant Cell*, **22**, 606-622.

Bhattacharya, S. and Baldwin, I.T. (2012) The post-pollination ethylene burst and the continuation of floral advertisement are harbingers of non-random mate selection in *Nicotiana attenuata*. *Plant J*, **71**, 587-601.

Bunning, E. (1956) Endogenous rhythms in plants. *Annu Rev Plant Physiol*, **7**, 71-90.

Corellou, F., Schwartz, C., Motta, J.-P., Djouani-Tahri, E.B., Sanchez, F. and Bouget, F.-Y. (2009) Clocks in the green lineage: comparative functional analysis of the circadian architecture of the *Picoeukaryote Ostreococcus*. *Plant Cell*, **21**, 3436-3449.

Edgar, R.S., Green, E.W., Zhao, Y., van Ooijen, G., Olmedo, M., Qin, X., Xu, Y., Pan, M., Valekunja, U.K., Feeney, K.A., Maywood, E.S., Hastings, M.H., Baliga, N.S., Merrow, M., Millar, A.J., Johnson, C.H., Kyriacou, C.P., O'Neill, J.S. and Reddy, A.B. (2012) Peroxiredoxins are conserved markers of circadian rhythms. *Nature*, **485**, 459-464.

Filichkin, S.A., Breton, G., Priest, H.D., Dharmawardhana, P., Jaiswal, P., Fox, S.E., Michael, T.P., Chory, J., Kay, S.A. and Mockler, T.C. (2011) Global profiling of rice and poplar transcriptomes highlights key conserved circadian-controlled pathways and *cis*-regulatory modules. *PLoS One*, **6**, e16907.

Fründ, J., Dormann, C.F. and Tscharntke, T. (2011) Linné's floral clock is slow without pollinators – flower closure and plant-pollinator interaction webs. *Ecol Lett*, **14**, 896-904.

Gase, K., Weinhold, A., Bozorov, T., Schuck, S. and Baldwin, I.T. (2011) Efficient screening of transgenic plant lines for ecological research. *Molecular ecology resources*, **11**, 890-902.

Gendron, J.M., Pruneda-Paz, J.L., Doherty, C.J., Gross, A.M., Kang, S.E. and Kay, S.A. (2012) Arabidopsis circadian clock protein, TOC1, is a DNA-binding transcription factor. *P Natl Acad Sci USA*, **109**, 3167-3172.

Goodspeed, D., Chehab, E.W., Min-Venditti, A., Braam, J. and Covington, M.F. (2012) *Arabidopsis* synchronizes jasmonate-mediated defense with insect circadian behavior. *P Natl Acad Sci USA*, **109**, 4674-4677.

Hsu, P.Y., Devisetty, U.K. and Harmer, S.L. (2013) Accurate timekeeping is controlled by a cycling activator in Arabidopsis. *eLife Sciences*, **2**, e00473.

Huang, W., Pérez-García, P., Pokhilko, A., Millar, A.J., Antoshechkin, I., Riechmann, J.L. and Mas, P. (2012) Mapping the core of the

Arabidopsis circadian clock defines the network structure of the oscillator. *Science*, **336**, 75-79.

James, A.B., Monreal, J.A., Nimmo, G.A., Kelly, C.L., Herzyk, P., Jenkins, G.I. and Nimmo, H.G. (2008) The circadian clock in *Arabidopsis* roots is a simplified slave version of the clock in shoots. *Science*, **322**, 1832-1835.

Kessler, D. and Baldwin, I.T. (2007) Making sense of nectar scents: the effects of nectar secondary metabolites on floral visitors of *Nicotiana attenuata*. *Plant J*, **49**, 840-854.

Kessler, D., Diezel, C. and Baldwin, I.T. (2010) Changing pollinators as a means of escaping herbivores. *Curr Biol*, **20**, 237-242.

Khan, S., Rowe, S. and Harmon, F. (2010) Coordination of the maize transcriptome by a conserved circadian clock. *BMC Plant Biol*, **10**, 126.

Kiba, T., Henriques, R., Sakakibara, H. and Chua, N.-H. (2007) Targeted degradation of PSEUDO-RESPONSE REGULATOR5 by an SCFZTL complex regulates clock function and photomorphogenesis in *Arabidopsis thaliana*. *Plant Cell*, **19**, 2516-2530.

Kim, W.Y., Fujiwara, S., Suh, S.S., Kim, J., Kim, Y., Han, L.Q., David, K., Putterill, J., Nam, H.G. and Somers, D.E. (2007) ZEITLUPE is a circadian photoreceptor stabilized by GIGANTEA in blue light. *Nature*, **449**, 356-+.

Kolosova, N., Gorenstein, N., Kish, C.M. and Dudareva, N. (2001) Regulation of circadian methyl benzoate emission in diurnally and nocturnally emitting plants. *Plant Cell*, **13**, 2333-2347.

Krügel, T., Lim, M., Gase, K., Halitschke, R. and Baldwin, I.T. (2002) *Agrobacterium*-mediated transformation of *Nicotiana attenuata*, a model ecological expression system. *Chemoecology*, **12**, 177-183.

Liu, H., Wang, H., Gao, P., Xu, J., Xu, T., Wang, J., Wang, B., Lin, C. and Fu, Y.F. (2009) Analysis of clock gene homologs using unifoliolates as target organs in soybean (*Glycine max*). *Journal of plant physiology*, **166**, 278-289.

Lou, P., Wu, J., Cheng, F., Cressman, L.G., Wang, X. and McClung, C.R. (2012) Preferential retention of circadian clock genes during diploidization following whole genome triplication in *Brassica rapa*. *Plant Cell*, **24**, 2415–2426.

Loughrin, J.H., Hamilton-Kemp, T.R., Andersen, R.A. and Hildebrand, D.F. (1991) Circadian rhythm of volatile emission from flowers of *Nicotiana sylvestris* and *N. suaveolens*. *Physiologia Plantarum*, **83**, 492-496.

Más, P., Alabadí, D., Yanovsky, M.J., Oyama, T. and Kay, S.A. (2003) Dual role of TOC1 in the control of circadian and photomorphogenic responses in Arabidopsis. *Plant Cell*, **15**, 223-236.

Mas, P., Kim, W.-Y., Somers, D.E. and Kay, S.A. (2003) Targeted degradation of TOC1 by ZTL modulates circadian function in *Arabidopsis thaliana*. *Nature*, **426**, 567-570.

McClung, C.R. (2006) Plant circadian rhythms. *Plant Cell*, **18**, 792-803.

McClung, C.R. (2013) Beyond Arabidopsis: The circadian clock in non-model plant species. *Seminars in Cell & Developmental Biology*, **24**, 430-436.

Mizoguchi, T., Wheatley, K., Hanzawa, Y., Wright, L., Mizoguchi, M., Song, H.-R., Carré, I.A. and Coupland, G. (2002) LHY and CCA1 are partially redundant genes required to maintain circadian rhythms in Arabidopsis. *Developmental Cell*, **2**, 629-641.

Murakami, M., Tago, Y., Yamashino, T. and Mizuno, T. (2007) Comparative overviews of clock-associated genes of *Arabidopsis thaliana* and *Oryza sativa*. *Plant and Cell Physiology*, **48**, 110-121.

Nagel, Dawn H. and Kay, Steve A. (2012) Complexity in the wiring and regulation of plant circadian networks. *Curr Biol*, **22**, R648-R657.

Nelson, D.C., Lasswell, J., Rogg, L.E., Cohen, M.A. and Bartel, B. (2000) FKF1, a clock-controlled gene that regulates the transition to flowering in *Arabidopsis*. *Cell*, **101**, 331-340.

Nitta, K., Yasumoto, A.A. and Yahara, T. (2010) Variation of flower opening and closing times in F1 and F2 hybrids of daylily (*Hemerocallis Fulva*; Hemerocallidaceae) and nightlily (*H. Citrina*). *Am J Bot*, **97**, 261-267.

Niwa, Y., Yamashino, T. and Mizuno, T. (2009) The circadian clock regulates the photoperiodic response of hypocotyl elongation through a coincidence mechanism in *Arabidopsis thaliana*. *Plant and Cell Physiology*, **50**, 838.

O'Neill, J.S. and Reddy, A.B. (2011) Circadian clocks in human red blood cells. *Nature*, **469**, 498-503.

Overland, L. (1960) Endogenous rhythm in opening and odor of flowers of *Cestrum nocturnum*. *Am J Bot*, **47**, 378-382.

Potts, S.G., Biesmeijer, J.C., Kremen, C., Neumann, P., Schweiger, O. and Kunin, W.E. (2010) Global pollinator declines: trends, impacts and drivers. *Trends Ecol Evol*, **25**, 345-353.

Raguso, R.A. (2004) Flowers as sensory billboards: progress towards an integrated understanding of floral advertisement. *Curr Opin Plant Biol*, **7**, 434-440.

Ramos, A., Pérez-Solís, E., Ibáñez, C., Casado, R., Collada, C., Gómez, L., Aragoncillo, C. and Allona, I. (2005) Winter disruption of the circadian clock in chestnut. *P Natl Acad Sci USA*, **102**, 7037-7042.

Sawa, M., Nusinow, D.A., Kay, S.A. and Imaizumi, T. (2007) FKF1 and GIGANTEA complex formation is required for day-length measurement in *Arabidopsis*. *Science*, **318**, 261-265.

Schaffer, R., Ramsay, N., Samach, A., Corden, S., Putterill, J., Carré, I.A. and Coupland, G. (1998) The late elongated hypocotyl mutation of *Arabidopsis* disrupts circadian rhythms and the photoperiodic control of flowering. *Cell*, **93**, 1219-1229.

Seo, P.J., Park, M.-J., Lim, M.-H., Kim, S.-G., Lee, M., Baldwin, I.T. and Park, C.-M. (2012) A self-regulatory circuit of CIRCADIAN CLOCK-ASSOCIATED1 underlies the circadian clock regulation of temperature responses in Arabidopsis. *Plant Cell.*

Somers, D.E. (1999) The physiology and molecular bases of the plant circadian clock. *Plant Physiol*, **121**, 9-20.

Somers, D.E., Kim, W.Y. and Geng, R. (2004) The F-box protein ZEITLUPE confers dosage-dependent control on the circadian clock, photomorphogenesis, and flowering time. *Plant Cell*, **16**, 769-782.

Somers, D.E., Schultz, T.F., Milnamow, M. and Kay, S.A. (2000) ZEITLUPE encodes a novel clock-associated PAS protein from Arabidopsis. *Cell*, **101**, 319-329.

Somers, D.E., Webb, A., Pearson, M. and Kay, S.A. (1998) The short-period mutant, *toc1-1*, alters circadian clock regulation of multiple outputs throughout development in *Arabidopsis thaliana. Development*, **125**, 485-494.

Staiger, D., Shin, J., Johansson, M. and Davis, S. (2013) The circadian clock goes genomic. *Genome Biology*, **14**, 208.

Strayer, C., Oyama, T., Schultz, T.F., Raman, R., Somers, D.E., Más, P., Panda, S., Kreps, J.A. and Kay, S.A. (2000) Cloning of the *Arabidopsis* clock gene *TOC1*, an autoregulatory response regulator homolog. *Science*, **289**, 768.

Sweeney, B.M. (1963) Biological clocks in plants. *Annu Rev Plant Physiol*, **14**, 411-440.

Takata, N., Saito, S., Tanaka Saito, C., Nanjo, T., Shinohara, K. and Uemura, M. (2009) Molecular phylogeny and expression of poplar circadian clock genes, *LHY1* and *LHY2*. *New Phytologist*, **181**, 808-819.

van Doorn, W.G. and Van Meeteren, U. (2003) Flower opening and closure: a review. *J Exp Bot*, **54**, 1801-1812.

Wang, W., Barnaby, J.Y., Tada, Y., Li, H., Tor, M., Caldelari, D., Lee, D.-u., Fu, X.-D. and Dong, X. (2011a) Timing of plant immune responses by a central circadian regulator. *Nature*, **470**, 110-114.

Wang, X., Wu, L., Zhang, S., Wu, L., Ku, L., Wei, X., Xie, L. and Chen, Y. (2011b) Robust expression and association of ZmCCA1 with circadian rhythms in maize. *Plant Cell Rep*, **30**, 1261-1272.

Xu, X., Xie, Q. and McClung, C.R. (2010) Robust circadian rhythms of gene expression in *Brassica rapa* tissue culture. *Plant Physiol*, **153**, 841-850.

Xue, Z.G., Zhang, X.M., Lei, C.F., Chen, X.J. and Fu, Y.F. (2011) Molecular cloning and functional analysis of one ZEITLUPE homolog GmZTL3 in soybean. *Molecular Biology Reports*, 1-8.

Yakir, E., Hilman, D., Harir, Y. and Green, R.M. (2007) Regulation of output from the plant circadian clock. *FEBS Journal*, **274**, 335-345.

Yang, R. and Su, Z. (2010) Analyzing circadian expression data by harmonic regression based on autoregressive spectral estimation. *Bioinformatics*, **26**, i168-i174.

Yon, F., Seo, P.J., Ryu, J.Y., Park, C.M., Baldwin, I.T. and Kim, S.G. (2012) Identification and characterization of circadian clock genes in a native tobacco, *Nicotiana attenuata*. *BMC Plant Biol*, **12**, 172.

Supporting information

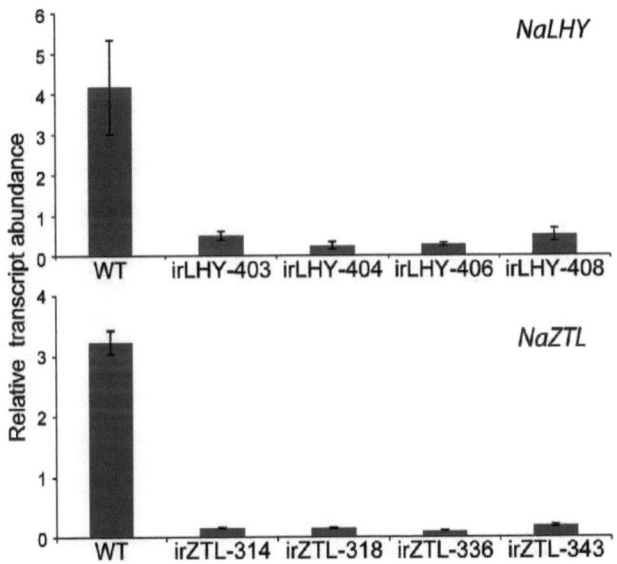

Figure S1. Silencing efficiency of irLHY and irZTL lines. Mean (± SE) levels of transcript accumulation of *NaLHY* and *NaZTL* in irLHY and irZTL lines, respectively. Plants were grown under 16 h light and 8 h dark conditions, and leaf samples were collected at ZT0 for irLHY, at ZT12 for irZTL lines. WT, wild-type; ZT, zeitgeber time.

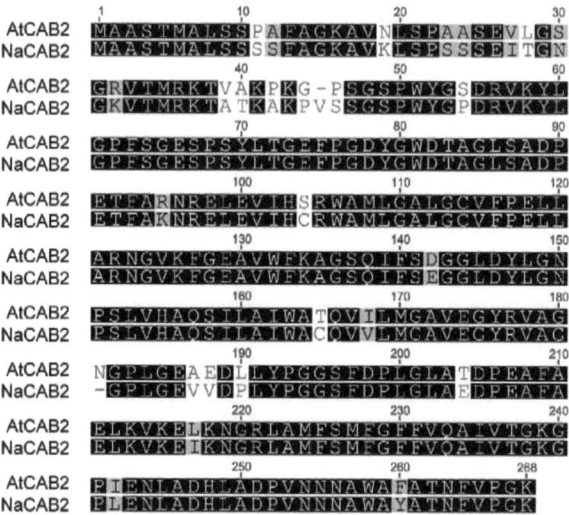

Figure S2. Protein alignment of CAB2 orthologs in *N. attenuata* and *A. thaliana*. Full-length amino acid sequences were aligned using the Geneious software V5.7.7 (www.geneious.com). TAIR accession number of *A. thaliana* CAB2 (AtCAB2) is AT1G29920. CAB2, CHLOROPHYLL A/B BINDING PROTEINS 2.

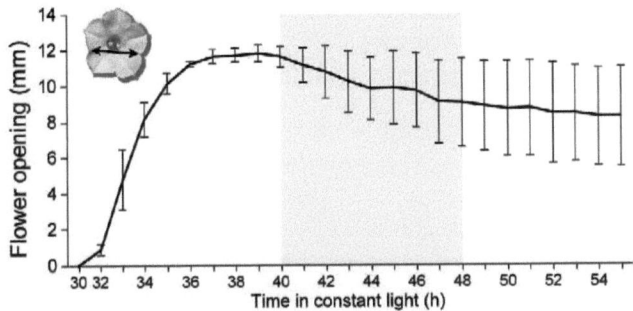

Figure S3. Flower opening and closing under constant light conditions. Mean (±SE) distance between petal junctions on corolla limbs of wild-type plants under constant light conditions. We exposed LD-grown flowering plants to LL condition for 24 h and measured flower opening/closing. A gray box indicates the subjective dark period of LL conditions.

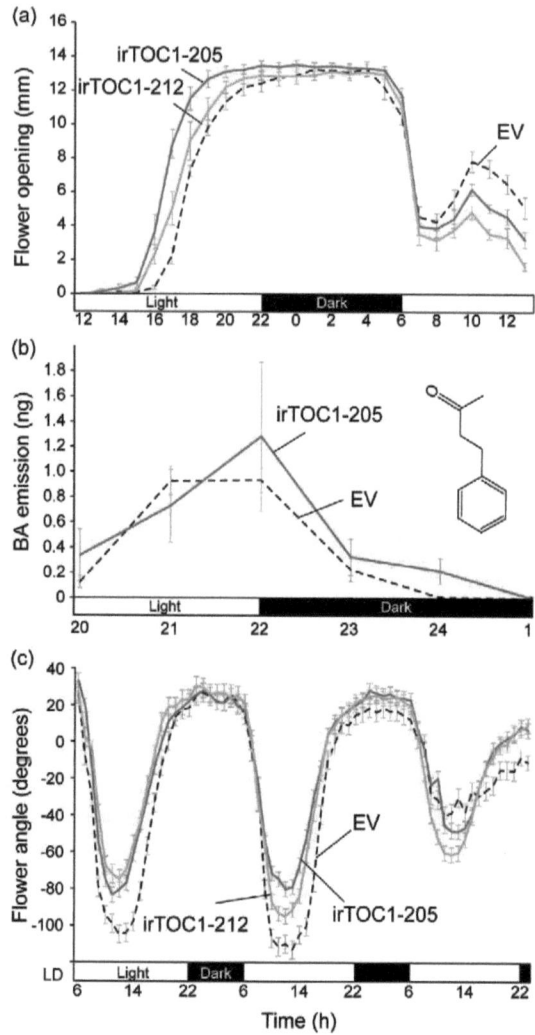

Figure S4. Floral phenotypes in irTOC1 lines. We measured flower opening, benzyl acetone (BA) emission, and flower angles of irTOC1 lines as described in Materials and Methods. Silencing efficiency of irTOC1 lines was shown in Yon et al.(25).

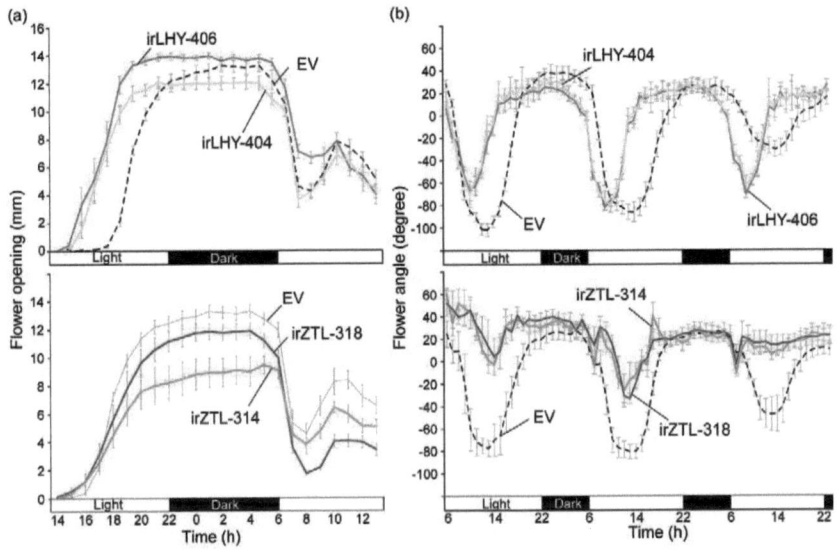

Figure S5. Floral phenotypes in two independent lines of irLHY and irZTL. We measured flower opening and flower angles of two independent lines of the clock gene-silenced lines: irLHY404, irLHY-406, irZTL-314, and irZTL-318, as described in Materials and Methods.

Chapter 5 – Manuscript III

Fitness consequences of altering circadian rhythms in *Nicotiana attenuata* flowers

Felipe Yon, Danny Kessler, Lucas Cortés Llorca, Youngsung Joo, Eva Rothe, Ian T. Baldwin and Sang-Gyu Kim

Department of Molecular Ecology, Max Planck Institute for Chemical Ecology, Hans-Knöll-Straße 8, D-07745 Jena, Germany.

Abstract

Ecological interactions between flowers and pollinators are all about timing. Flower opening/closing and scent emissions are largely synchronized with pollinators' active time, and a plant circadian clock regulates these rhythmic traits. Despite its importance for outcrossing, the hypothesis that the circadian clock increases the reproductive success by regulating floral rhythms is rarely tested. To test this, we examined the outcrossing success of the wild tobacco *Nicotiana attenuata* flowers, which rhythmically open, emit scents and move vertically to interact with nocturnal hawkmoths and also day-active pollinators, such as hummingbirds. Under both glasshouse and field conditions, we examined the outcrossing success of phase-shifting flowers generated by silencing circadian clock genes. The results demonstrated that the circadian rhythms in *N. attenuata* flowers influence the successful outcrossing and the choice of pollinators as well.

INTRODUCTION

In the 18th century, Carl Linnaeus noticed that many flowers open or close at specific times of a day, and he designed a garden known as the "flower clock" with these flowers (Somers 1999). Special diurnal rhythms in flowers, including opening/closing and scent emissions, have been evolved to synchronize with pollinators' active time for successful outcrossing (Fründ *et al.* 2011). For instance, the wild tobacco *Nicotiana attenuata*, which inhabits the Great Basin Desert in USA, produces two kinds of self-compatible flowers to interact with different pollinators: night-opening flowers for nocturnal hawkmoths (e.g. *Manduca sexta* and *M. quinquemaculata*) and morning-opening flowers for day-active pollinators such as hummingbirds (Kessler *et al.* 2010). Approximately 90% of *N. attenuata* flowers opens at night and emits benzyl acetone (BA), the main floral volatile compound that attracts nocturnal hawkmoths (Kessler *et al.* 2008), and the flowers close by the next morning (Kessler *et al.* 2010). These floral rhythms are repeated for two or three days. The relatively small number of flowers partially opens in the morning with reduced BA emissions and fully opens in the next night (Kessler *et al.* 2010). In addition, *N. attenuata* flowers show a special diurnal rhythm: flowers face downward in the morning and upward during the night.

To examine the ecological relevance of the diurnal rhythms in flowers, the choice of a proper model system with the physical or genetic manipulation of floral rhythms is essential (Resco *et al.* 2009). Several experiments under constant conditions suggested that an internal clock, which is called a circadian clock, regulates diurnal rhythms in flowers (Sweeney 1962; Hoballah *et al.* 2005). The plant circadian clock has been identified in a model plant species, *Arabidopsis thaliana* (Nagel & Kay 2012), and these clock components are highly conserved in many plant species (McClung 2013). The two morning elements of Arabidopsis clock, *LATE ELONGATED HYPOCOTYLE* (*LHY*) and

CIRCADIAN CLOCK ASSOCIATED1 (*CCA1*) negatively regulate the transcription of the evening element, *TIMING OF CAB EXPRESSION1* (*TOC1*). In the evening, induced TOC1 protein inhibits *LHY* and *CCA1* expressions, forming the core negative feedback loop of the clock. Another clock component, ZEITLUPE (ZTL) physically binds and degrades TOC1 protein in a light dependent manner (Kim *et al.* 2007) (Fig. S1). Altering the expression of the clock genes produces arrhythmic or dysrhythmic plants, which show many developmental (Adams & Carré 2011) and metabolic (Wang *et al.* 2011a; Goodspeed *et al.* 2012) defects. However, to our knowledge, there is no report of which clock components regulate floral rhythms. Moreover, the most of the clock-altered plants were generated from the species which show little interaction between its flowers and pollinators.

Previously we identified the homologous genes of the core clock components, *LHY*, *TOC1*, and *ZTL* in *N. attenuata* (Yon *et al.* 2012). To manipulate diurnal rhythms in flower, we silenced these clock genes in *N. attenuata* by transformation with gene-specific inverted-repeat (ir) constructs and found that silencing the clock genes alters floral rhythms in *N. attenuata* (Fig. 1)(Yon *et al.* 2012). The first objective of this study was to demonstrate the effects of altering floral rhythms for outcrossing mediated by *M. sexta*. The second was to test advantages and/or disadvantages of altering floral rhythms under field conditions, where several simultaneous interactions shape different fitness outcomes. We assessed the fitness by measuring the capsules and seed set produced from the emasculated flowers. Our results demonstrated that altering floral traits results in the change of the outcrossing rate in *N. attenuata* and the change of pollinators.

MATERIAL AND METHODS

Plant growth conditions

We used *Nicotiana attenuata* Torr. Ex. Wats (Solanaceae) plants (30st inbred generation) and isogenic silenced transformed plants, originated from a population in Utah. Seeds were sterilized and germinated on petri dishes and kept under long-day conditions (LD, 16h light/ 8h dark) in a growth chamber for 10 days until transferred to pots in a glasshouse as described in Krügel *et al.* 2002.

In the field, fifteen days old seedlings were transferred into hydrated 50-mm peat pellets until adapted to the environmental conditions in the Great Basin Desert. And later we transplanted them to an irrigated plot at the Lytle Ranch Preserve as described in Kessler *et al.* 2012. The release of transgenic plants was carried out under Animal and Plant Health Inspection Service releases 11-350-101r and 12-333-101r.

Floral traits of the clock-gene silenced lines in *N. attenuata*

NaLHY (NCBI accession number JQ424913), NaTOC1 (JQ424914) and NaZTL (JQ424912) was independently silenced by the transformation of gene-specific inverted repeat (ir) construct driven by the CaMV 35S promoter (Yon *et al.* 2012). Two independent T2 and T3 homozygous lines (irLHY404, irLHY-406, irTOC1-205, irTOC1-212, irZTL-314, and irZTL-318) were used to characterize diurnal rhythms in flowers (Yon *et al.* unpublished data) and all pollination results of this study were derived from irLHY-406, irTOC1-205, and irZTL-314. All floral traits of the clock-altered lines shown in Fig. 1 were summary of the result from our previous work (Yon *et al.* unpublished data).

Cross pollination experiment

To measure outcrossing rates in the clock-silenced lines, plants with emasculated flowers were transferred to a table covered with a green mesh tent of 1.8m height x 1.6m width x 6m length in a glasshouse. Fully developed flowers in LD-grown plants were emasculated in the early morning to avoid self-pollination. We chose experimental days when there were no other flowering *N. attenuata* plants in the same glasshouse cabin except from our experimental plants. For the no-competition experiment, five flowers on five plants per each line (EV, irLHY, irTOC1, and irZTL) were emasculated in the morning and exposed to two *M. sexta* moths with 10 WT plants as a pollen donor for one night. For the paired-competition experiments, five flowers on four EV plants and five flowers on four of each clock-silenced line (irLHY, irTOC1, and irZTL) were emasculated in the morning and competed for the pollination services of two *M. sexta* moths with 10 WT pollen donor plants for one night. EV and clock-silenced plants were arranged in a pair of one plant per genotype with 30 cm apart. The number of matured capsules and seeds were counted after ripening.

Field pollination experiments were conducted in June 2012 and June 2013, on the field plot at the Lytle Ranch Preserve (Santa Clara, Utah, USA). Flowers were emasculated before 9 am to prevent self-pollination. In the 2012 season flowers were enclosed in mesh ventilated plastic cups and released at next morning after night opening, to be accessible to day-time pollinators (mainly hummingbirds). At dusk they were reclosed in their plastic cups to exclude the night-time pollinators. In the 2013 season punctured plastic bags fixed on top of bamboo sticks by a ring wire were used for covering the flowers in the same time periods. Additionally in the 2013 season night-time pollination experiments were conducted, on which the flowers were covered after emasculation, released at dusk for an entire night, and prior dawn they were

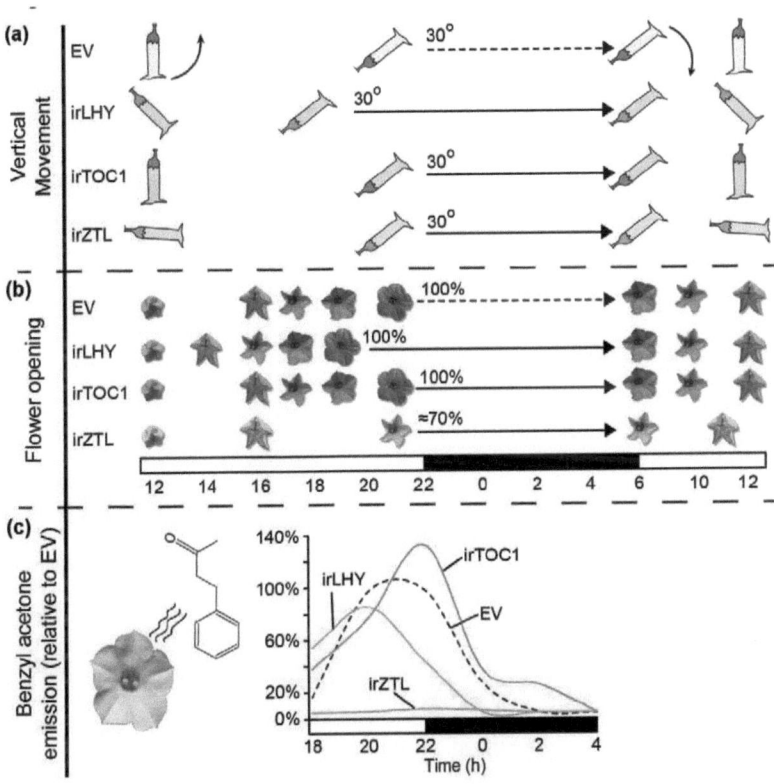

Figure 1. The circadian clock regulates floral rhythms in *N. attenuata*. Simple representation of (a) vertical movement and (b) aperture of the clock gene-silenced flowers in the first opening day. (c) Benzyl acetone (BA) emission trends in relative percentage to the maximum amount of BA emission from EV flowers. Inset figure depicts BA molecule. Each color indicates each transformed line. All plants were grown under long day conditions (16 h light:8 h dark). LHY, LATE ELONGATED HYPOCOTYL; TOC1, TIMING OF CAB EXPRESSION1; ZTL, ZEITLUPE; EV, plant transformed with the empty-vector used to generate transgenic lines; irLHY, NaLHY-silenced line; irTOC1, NaTOC1-silenced line; irZTL, NaZTL-silenced line.

covered to exclude day-time pollinators. Target plants were surrounded by WT as pollen donors, to ensure pollen flow in the experimental population. Formed capsules were counted and removed prior ripening to avoid seed release.

Statistical analysis

The results from capsule number and seed number were statistically analyzed using student's t-test and one-way ANOVA followed by Tukey-HSD post hoc tests, all were performed using R 2.15.3 (http://www.r-project.org/).

Figure 2. *Manduca sexta* hawkmoth approaching, probing and foraging on *N. attenuata* EV flowers facing naturally upwards. Photo sequence taken in glasshouse conditions at ca. 22 h with a wild-camera Snapshot Mini (Dörr, Germany), equipped with a PIR sensor camera and IR flash.

RESULTS

N. attenuata flowers maintain an approximately 40° upward position from the horizontal during the development (Fig. 1). In the morning of the first opening day, flowers move downward to approximately 90° below the horizontal; these flowers return to the upright position before dusk, and open (Fig. 1). By the next morning, flowers face down again and close their corollas. This vertical movement of flowers is repeated for 2-3 days under long day (LD, 16h light and 8h dark) conditions, with diminishing amplitude on the third day.

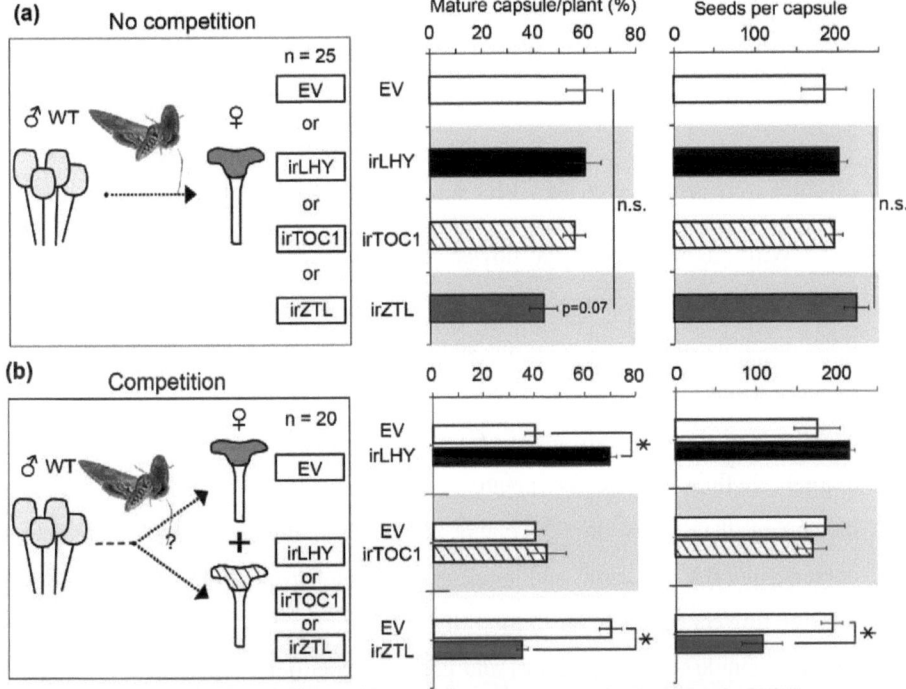

Figure 3. The circadian clock coordinates outcrossing success in *N. attenuata*. (a) Flowers of EV, irLHY, irTOC1, and irZTL lines were emasculated and exposed to two *M. sexta* moths for one night with 10 WT pollen donor plants. (b) In the competition experiments, one of the clock gene-silenced lines were paired with EV and competed for the pollination services of two *M. sexta* moths and 10 WT pollen donor plants. Mean (±SE) percentage of mature capsules per plant and mean (±SE) number of seeds per capsule resulting from outcrossing by *M. sexta* moths in emasculated flowers. Asterisks represent significant difference between EV and clock gene-silenced lines (** = $P < 0.01$, Student's *t*-test).

Why do *N. attenuata* flowers show diurnal vertical movement? Both the flight style and the construction of the proboscis of *M. sexta* moths (Fig. 2) (Sprayberry & Suver 2011), a main pollinator of *N. attenuata*, may restrict how the moth accesses the nectar reward of the flowers, and so we previously hypothesized that flower positions affect the success of *N. attenuata*'s outcrossing mediated by *M. sexta*. We previously tested this hypothesis by emasculating flowers to prevent self-pollination and fixed them at one of the three different positions (45°, 0° and -45°) and allowed *M. sexta* visitation. We found that the emasculated flowers tethered at 45° and 0° produced 65% and 35% capsules, respectively, from the total of emasculated flowers, and no capsules were produced at -45° (Llorca *et al*, unpublished data).

With the evidence of the flower orientation importance, in this study we used the clock gene-silenced plants to evaluate the ecological significance of floral rhythms in a fine resolution and under more natural conditions. In our previous study, we found that silencing circadian clock genes in *N. attenuata* alters diurnal rhythms in flowers; *LHY*-silenced (irLHY) plants show 2 h earlier diurnal rhythms in flower opening, BA emission, and upward vertical movement than those in EV plants, and irZTL flowers show incomplete opening, no BA emission, and weak vertical movement (Fig. 1). In contrast, irTOC1 flowers have similar diurnal rhythms compared to EV flowers (Fig. 1). We used single (no competition) and pairs (competition with EV) of plants in pollen-acceptor experiments (Fig. 3). For the no-competition experiment using a single line, we emasculated a total of 25 flowers (5 flowers/plant) on EV, irLHY, irTOC1, or irZTL plants and placed plants of a single line in the tent with two *M. sexta* moths and 10 WT pollen-donor plants for one night (Fig. 3A). After two weeks, 60% of EV flowers had produced mature capsules, and similar outcrossing rates were observed in irLHY and irTOC1, but irZTL flowers tended to produce fewer capsules than did flowers in EV ($t = 1.84$, $P = 0.07$). Seed numbers per capsule did not differ among the lines (Fig. 3A).

However, when pollinators were given a choice between visiting EV and clock-silenced lines, the rates of outcrossing success differed significantly. irZTL plants produced fewer capsules than EV plants (Fig. 3B, $t = 7.82$, $P < 0.001$). Furthermore, difference in capsule numbers of the EV-irZTL pair (Fig. 3B) was larger than the difference in capsule numbers between single EV and irZTL plants (Fig. 3A). EV plants in the EV-irZTL pair produced more capsules than EV plants in EV-irLHY or EV-irTOC1 pairs. In addition, irZTL line produced significantly fewer seeds per capsule (t = 3.42, $P < 0.05$) than did EV lines in EV-irZTL pairs (Fig. 3B). EV- irTOC1 pairs showed no difference either in capsule formation or in number of seeds per capsule (Fig. 3B). Unexpectedly, in the EV-irLHY pair, irLHY line produced a significantly larger number of matured capsules (t = 7.5, $P < 0.001$) than did EV, which was the opposite outcome compared with the EV-irZTL pair (Fig. 3B). While irLHY produced more capsules under competition with EV, there was no significant difference in the numbers of seeds per capsule in the EV- irLHY pair (t = 1.28, $P = 0.21$).

N. attenuata flowers are also visited by day-active pollinators, mainly hummingbirds or minor one of hymenoptera species. To examine whether the changes of floral rhythms affect the interaction with day-active pollinators, we performed outcrossing experiments under competition conditions over two successive field seasons in 2012 and 2013 in the native habitat of *N. attenuata*. Interestingly, the results showed the opposite trend compared to the tent experiment done with *M. sexta* moths. In 2012 field season, irZTL flowers produced significantly more matured capsules than did EV (t = 3.26, $P < 0.01$) and irLHY (t = 2.71, $P < 0.05$) flowers; 0%, 5%, and 30% of total emasculated flowers were matured in EV, irLHY, and irZTL plants, respectively (Fig. 4A). In the 2013 season, irZTL plants produced approximately 20% more capsules than EV and irLHY plants, but there was no significant difference among the lines ($F = 1.29$, $P = 0.3$) or in paired comparisons (EV-irZTL, t = 1.94, $P = 0.08$; EV-irLHY, t = 0.9, $P = 0.39$) (Fig. 4B).

We were able to perform pollination experiment with night-active pollinators in the 2013 field season because of the presence and positive identification of *M. sexta* moths. In contrast to the result from the tent experiment, EV plants produced twice more capsules than did irLHY plants (Fig. 4C). EV plants also produced 30% more capsules than irZTL plants (Fig. 4C), while irZTL flowers opened completely and emitted low amount of BA in the field conditions (unpublished data). But there was no significant difference among the lines (F = 0.87, P = 0.44) or in paired comparison (EV-irLHY, t = 1.28, P = 0.23; EV-irZTL, t = 0.65, P = 0.53).

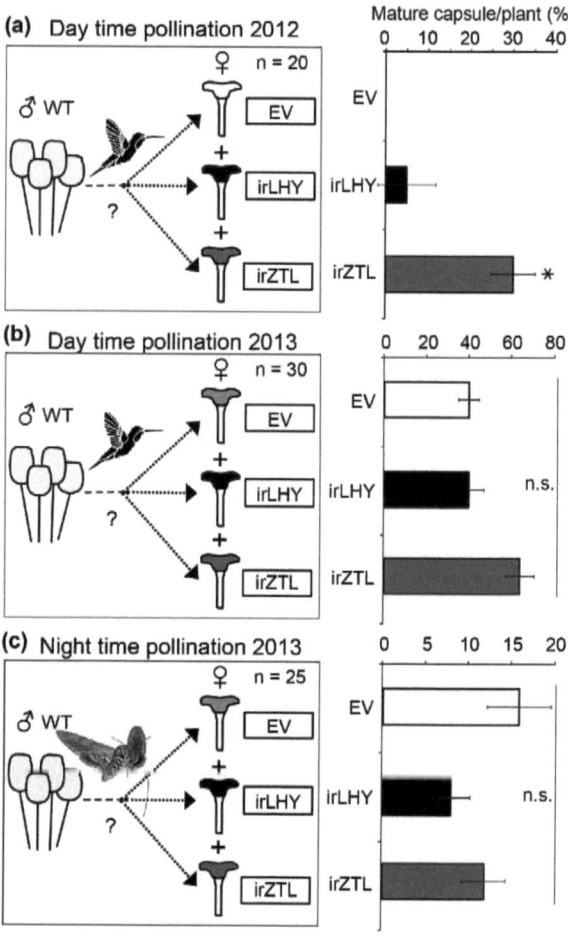

Figure 4. Alteration of the circadian clock confers time shift advantage in nature. (a) Flowers of EV, irLHY and irZTL lines were emasculated and exposed during day time after first opening, surrounded by WT pollen donors. Mean (±SE) percentage of mature capsules/plant and mean (±SE) number of seeds/capsule from outcrossing by *M. sexta*. Asterisks represent significant difference between EV and clock gene-silenced lines (** = P <0.01, *Chi* square test of independence).

DISCUSSION

Ecological implication of the circadian clock

The fundamental concept in chronobiology is that the circadian clock increases the fitness of organisms. Therefore, we predicted that disrhythmic/arrhythmic traits in the clock-altered flowers reduce the outcrossing success. As expected, outcrossing rates in irZTL plants was reduced when irZTL plants competed with EV plants to attract *M. sexta* moths in a tent (Fig. 3B). However, irLHY flowers had higher outcrossing rates when they competed with EV flowers in a tent (Fig. 3B), while their flower angles, opening, and the amount of BA emission were similar to those in EV flowers when *M. sexta* moths were most active (22 - 23 h) under our glasshouse conditions. The possible explanation is that floral volatiles over irLHY plants can be accumulated more than EV plants due to the 2 h earlier BA emission from irLHY flowers and low airflow conditions in a glasshouse. Alternative hypothesis is that other unmeasured floral traits, such as minor floral scents, green leaf volatiles, nectar volume or non-visible UV floral pigmentation, may be altered in irLHY flowers. However, this benefit of irLHY flowers was lost in the field (Fig. 4C). Although there is no statistical difference between irLHY and EV plants in the 2013 field experiment, irLHY flowers produced 50% less capsules than EV flowers. These results suggest that the early "advertisements behavior" in irLHY may increase the visitation of unfavorable insects in nature, such as florivores or nectar robbers, which reduce outcrossing success (Kessler & Baldwin 2011).

Interestingly, day-active pollinators visited irZTL flowers more than EV flowers in nature, suggesting that the circadian clock regulates floral rhythms as a means of selecting pollinators. This could be attained from the upward position of irZTL flowers in a day when EV flowers face downward. White corolla limb facing upward may attract more day-active pollinators, such as hummingbirds, which normally use visual cue to find flowers. Hymenoptera species were observed visiting flowers to collect pollen in the field, but they were not considered important pollinators in this experiment because we emasculated flowers for checking the outcrossing success. This result from the day-time experiment supports the hypothesis that visual stimuli are more important for day-time pollinators (Aizen 2003; Kessler *et al.* 2010; Clarke *et al.* 2013) in order to find *N. attenuata* flowers, and a downward orientation is an effective mean to reduce the visitation of unwanted pollinators (Fulton & Hodges 1999; Hodges *et al.* 2004).

Ecological implication of flower orientations

Several hypotheses have been formulated about the effect of flower orientation on outcrossing. For example, the horizontal or downward orientations of flowers increase pollen transfer because pollinators are in contact with flowers for a longer time (Fenster *et al.* 2009). In addition, the upward orientation of flowers facilitates a multi-directional recognition by pollinators despite the reduction of pollen transfer (Ushimaru & Hyodo 2005; Fenster *et al.* 2009). Why down? Downward orientation of flowers gives several advantages: it reduces susceptibility to florivores (Ashman & Schoen 1994) and nectar desiccation caused by solar radiation(Kessler 2012), and it excludes the visit of day-active pollinators (Fenster *et al.* 2004). Sugar concentrations of the nectar in *N. attenuata* flowers are decreased by the strong sun light in the Great Basin Desert (Climate 101 2013), although flowers face downward and close during the day (Kessler 2012). If flowers face upward and open during the day, this effect could be greater. Other literature shows that the downward orientation of flowers helps to avoid pollen wash and nectar dilution due to rain exposition (Tadey & Aizen 2001; Aizen 2003). However, nectar robbing carpenter bees can collect nectar by puncturing the corolla tube at dawn and dusk (Kessler *et al.* 2008), suggesting that the vertical movement alone cannot prevent some degree of damages by opportunistic robbers.

Circadian rhythms in flowers have been developed to interact with mutualist and antagonist and also to synchronize with the environmental rhythms in their native habitats. The main purpose is to ensure outcrossing services by attracting pollinators in a right time (Jones & Little 1983; Harder & Barrett 2006). *N. attenuata* is an interesting model species to study the function of floral rhythms for plant-pollinator interactions, because *N. attenuata* produces two kinds of flowers which show three diurnal rhythms to attract the different types of pollinators (Kessler *et al.* 2010). By manipulating floral rhythms genetically, we clearly show that altering circadian rhythms in flowers affect outcrossing success under both lab and field conditions and also the choice of pollinators.

ACKNOWLEDGEMENTS

We thank Dr. Klaus Gase for designing ir-constructs, and the Brigham Young University for use of their field station, the Lytle Ranch Preserve. All authors declare that they have no conflicts of interest. This work is supported by European Research Council advanced grant ClockworkGreen (No. 293926) to ITB, the Global Research Lab program (2012055546) from the National Research Foundation of Korea, and the Max Planck Society.

REFERENCES

1.

Adams, S. & Carré, I.A. (2011). Downstream of the plant circadian clock: output pathways for the control of physiology and development. *Essays in Biochemistry*, 49, 53–69.

2.

Aizen, M.A. (2003). Down-Facing Flowers, Hummingbirds and Rain. *Taxon*, 52, 675.

3.

Ashman, T.-L. & Schoen, D.J. (1994). How long should flowers live? *Nature*, 371.

4.

Clarke, D., Whitney, H., Sutton, G. & Robert, D. (2013). Detection and learning of floral electric fields by bumblebees. *Science*, 340, 66–9.

5.

Climate 101. (2013). Climate 101 [WWW Document]. URL http://www.climate101.org/2013/03/the-great-basin/.

6.

Fenster, C.B., Armbruster, W.S. & Dudash, M.R. (2009). Specialization of flowers: is floral orientation an overlooked first step? *The New phytologist*, 183, 502–6.

7.

Fenster, C.B., Armbruster, W.S., Wilson, P., Dudash, M.R. & Thomson, J.D. (2004). Pollination Syndromes and Floral Specialization. *Annual Review of Ecology, Evolution, and Systematics*, 35, 375–403.

8.

Fründ, J., Dormann, C.F. & Tscharntke, T. (2011). Linné's floral clock is slow without pollinators - flower closure and plant-pollinator interaction webs. *Ecology letters*, 14, 896–904.

9.

Fulton, M. & Hodges, S.A. (1999). Floral isolation between Aquilegia formosa and Aquilegia pubescens. *Proceedings of the Royal Society B-Biological Sciences*, 266.

10.

Goodspeed, D., Chehab, E.W., Min-Venditti, A., Braam, J. & Covington, M.F. (2012). Arabidopsis synchronizes jasmonate-mediated defense with insect circadian behavior. *Proceedings of the National Academy of Sciences*, 109, 4674–4677.

11.

Harder, L.D. & Barrett, S.C.H. (2006). *Ecology and Evolution of Flowers*. Oxford University Press, New York.

12.

Hoballah, M.E., Stuurman, J., Turlings, T.C.J., Guerin, P.M., Connétable, S. & Kuhlemeier, C. (2005). The composition and timing of flower odour emission by wild Petunia axillaris coincide with the antennal perception and nocturnal activity of the pollinator Manduca sexta. *Planta*, 222, 141–50.

13.

Hodges, S.A., Fulton, M., Yang, J.Y. & Whittall, J.B. (2004). Verne Grant and evolutionary studies of Aquilegia. *New Phytologist*, 161, 113–120.

14.

Jones, C.E. & Little, R.J. (1983). *Handbook of experimental pollination biology*. Van Nostrand Reinhold, New York.

15.

Kessler, D. (2012). Context dependency of nectar reward-guided oviposition. *Entomologia Experimentalis et Applicata*, 144, 112–122.

16.

Kessler, D. & Baldwin, I.T. (2011). Back to the past for pollination biology. *Current opinion in plant biology*, 14, 429–34.

17.

Kessler, D., Diezel, C. & Baldwin, I.T. (2010). Changing pollinators as a means of escaping herbivores. *Current Biology*, 20, 237–42.

18.

Kessler, D., Gase, K. & Baldwin, I.T. (2008). Field experiments with transformed plants reveal the sense of floral scents. *Science*, 321, 1200–2.

19.

Kim, W.-Y., Fujiwara, S., Suh, S.-S., Kim, J., Kim, Y., Han, L., *et al.* (2007). ZEITLUPE is a circadian photoreceptor stabilized by GIGANTEA in blue light. *Nature*, 449, 356–60.

20.

McClung, C.R. (2013). Beyond Arabidopsis: The circadian clock in non-model plant species. *Seminars in cell & developmental biology*, 24, 430–436.

21.

Nagel, D.H. & Kay, S. a. (2012). Complexity in the wiring and regulation of plant circadian networks. *Current biology : CB*, 22, R648–57.

22.

Resco, V., Hartwell, J. & Hall, A. (2009). Ecological implications of plants' ability to tell the time. *Ecology Letters*, 12, 583–592.

23.

Somers, D.E. (1999). The Physiology and Molecular Bases of the Plant Circadian Clock. *PLANT PHYSIOLOGY*, 121, 9–20.

24.

Sprayberry, J.D.H. & Suver, M. (2011). Hawkmoths' innate flower preferences: a potential selective force on floral biomechanics. *Arthropod-Plant Interactions*, 263–268.

25.

Sweeney, B.M. (1962). Biological clocks in plants. *Annual Review Plant Physiology*, 411–440.

26.

Tadey, M. & Aizen, M.A. (2001). Why do flowers of a hummingbird-pollinated mistletoe face down? *Functional Ecology*, 15, 782–790.

27.

Ushimaru, A. & Hyodo, F. (2005). Why do bilaterally symmetrical flowers orient vertically? Flower orientation influences pollinator landing behaviour. *Evolutionary Ecology Research*, 7, 151–160.

28.

Wang, W., Barnaby, J.Y., Tada, Y., Li, H., Tör, M., Caldelari, D., *et al.* (2011). Timing of plant immune responses by a central circadian regulator. *Nature*, 470, 110–4.

29.

Yon, F., Seo, P.J., Ryu, J.Y., Park, C.-M., Baldwin, I.T. & Kim, S.-G. (2012). Identification and characterization of circadian clock genes in a native tobacco, Nicotiana attenuata. *BMC plant biology*, 12, 172.

SUPPORTING INFORMATION

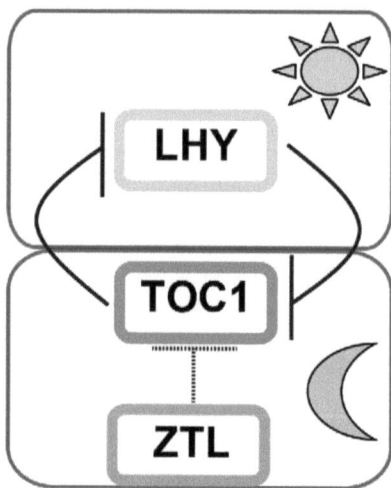

Figure S1. Simplified clock core components interaction. LHY and TOC1 genes inhibit each other transcription at different times of the day (solid black lines), whether ZTL proteins targets TOC1 proteins for degradation at night time (dashed black lines).

Chapter 5 – Manuscript III

Chapter 6 – Discussion

The daily rotation cycles of Earth had dictated the living rhythm of organisms, adapting to available times to acquire energy and nutrients, two processes not necessarily coupled on time in the photosynthetic organisms. The output of endogenous clocks has been observed with detail for some centuries. An example is the flower clock garden of Linnaeus. Ever since the deduction of circadian clock components and mechanism it has been studied in several organisms, with different findings, like animals that have a centralized pacemaker, unlike plants, or the case of unicellular organisms with more than a single internal clock (Roenneberg & Mittag 1996). Although several clock mutants have been identified in species of the plant kingdom, its use for understanding the reproduction dynamics that are established with animals have been largely ignored. This is of particular interest to conservation ecology and crops pollination given the actual deterioration context, where pollinator species and assemblages are lost at fast pace, in part to the climate pattern change that causes species and habitat shifts. The comprehension of how these dynamic interactions are regulated by the inner plant rhythms would be a strong asset to analyze poor reproduction and propagation of plant species in their original or new habitats. In this work, with the results from Manuscript II and III, the importance of the regulation and synchronization of the floral traits in order to interact with pollinators is presented and discussed.

Different clock in species

Since the first observations and descriptions of an endogenous clock in plants, considerable progress has been made in describing the components and mechanisms that make the circadian clock work in plants. The currently accepted clock model in plants was developed by extensive studies in the model plant *A. thaliana* due to its manipulation easiness and relatively small genome. Nowadays, the clock has been described in several photosynthetic species, sharing the same components at the higher plants taxa but not necessarily on earlier photosynthetic taxa (McClung 2013). Therefore, assuming a uniform circadian clock model will be misleading, already considering the diversity within the evolutionary Plantae kingdom.

A more suitable assumption will be to consider a variety of evolutionary circadian-clock models, which despite sharing features as closer they are in evolutionary time, will not necessarily share the same regulation and phenotypic output. This is evidenced by the number of clock components that are replicated in some species. For example, the CCA1-like gene is so far not commonly found in monocots, or the case of multiple LHY-like genes coincidences in poplar and chestnut (Ramos *et al.* 2005; Takata *et al.* 2009). Thus far CCA1/LHY like genes have not been found in taxa not belonging to the land plants. Given that, it can be assumed that it appeared after land colonization and then it duplicated to separate in one of each or both functional genes. Evidence can be found in the moss *Physcomitrella patens*, which presents two CCA1-copies (Okada *et al.* 2009), but not a defined LHY copy.

In the case of *N. attenuata,* as shown in Manuscript I, only one LHY gene copy was identified, and up to now not a copy of CCA1. The normal oscillation of the circadian clock in *N. attenuata* suggests that the presence of a CCA1 copy is not necessary for a functional endogenous clock oscillation. This normal clock function with only one of both genes can be explained by the fact that CCA1 and LHY have overlapping time keeping functions, as already shown in other organisms that work normally with one copy.

Given the different clock components composition in land plants and lower taxa, it is apparent to assume a differential clock output depending on the species group, where different pathways will have differential regulation in response to the plant metabolic needs. And additive factor in a differential clock output will be the environmental conditions on which the plant species thrive. Because if a species is present in a continuous habitat range, the environmental conditions change between localities imposing different resources and interactions, which imply a variety of time restrictions (Hut *et al.* 2013). This has been observed in cross latitudinal studies in *A. thaliana*. An example of this is the flowering time that changes between ecotypes, as each one is adapted to a particular photoperiod ranging from Mediterranean to circumpolar habitats (Hancock *et al.* 2011). Under this view it's possible to expect that *N. attenuata* accession originated from Utah has a different flowering time compare to accessions grown in other habitats.

Similar variation was demonstrated using clock mutants from *Arabidopsis* accessions that do not show delayed flowering, like the case of *toc1-1* mutant, where C24 present late flowering under long day conditions but Columbia and Landberg accessions do not(Somers *et al.* 1998). In Manuscript I it was shown that irTOC1 line has a delayed flowering time, but on different genetic backgrounds different results would be probable. It is expected that other *N. attenuata* accessions will have differential clock plasticity of flowering time and different clock mutant responses depending on their environmental conditions, like latitude or altitude, which affect the photoperiod and temperature adaptations.

Differential clock in tissues

Circadian clocks are not only transcriptional oscillations, but also metabolic, as it was shown with peroxiredoxins, which are highly conserved across species and works under different species clock models (Edgar et al. 2012). As it was shown in a previous study, *N. attenuata* has different oscillatory rhythms between the leaf and root tissues, where different sets of genes are circadian regulated, and the same is observed in metabolites rhythms (Kim et al. 2011). Different plant tissues require different regulations as each responds to specific purposes and functions, and the environmental conditions are also different if compared the above- and underground tissues. In the case of aboveground tissues, its purpose can be separated by organ types, and between vegetative and reproductive tissue. The first is in charge of producing photosynthates and give structural support. The second accounts for the seed production. The last involves a set of different tasks such as attracting pollinators and/or seed dispersal depending on the species.

As described in the results of Manuscript II, *N. attenuata* pedicel and corolla tissues keep the same peaking time of *NaLHY*, but the corolla tissue shows a slower *NaLHY* transcript decrease rate, which differentiates it from the pedicel, a green tissue, which has the same decrease profile as the leaf tissue. An evidence of tissue independence was obtained while performing a trial experiment, on which *N. attenuata* rosette leaf tissue was covered with aluminium foil, the stalk leaves were removed and only the flowers were left exposed to light. During the five-day experiment the floral traits of young and old flowers kept their circadian rhythms independently of the continuous dark conditions from the vegetative tissue. This can be possible because photoreceptors are located in all plant tissues, which perceive the light cycles and provide the gating signals to entrain the circadian clock *in situ*(Yakir et al. 2011; Wenden et al. 2012). As demonstrated by studies on *Arabidopsis,* where different tissues were entrained at different phases or free running conditions, leaves and stalks of a single plant could be differentially set even within a same organ, such as a leaf between its tip and base (Thain *et al.* 2002; Yakir *et al.* 2011).

Considering the circadian clock independence of the plant tissues, it is possible to consider that each of these clocks have slight changes to synchronize the metabolic functions related with their function. The flower opening and its related BA emission at dusk could be associated by a trigger involving proteins regulated by TOC1 with peak at subjective dusk and slow transcript level decrease. Thus, mathematical models and tissue specific microarray data will be required in the future to correlate and analyze this association. So far the time of opening and BA emission can be associated with TOC1 through the use of clock-silenced lines, where the alterations of TOC1 peaking time cause variations in the floral traits timing, as seen in Manuscript II.

Single circadian clock components effects

The research in *N. attenuata* has lead overtime to the development of a molecular biology toolbox implementing reverse genetics approaches, such as gene silencing by inverted repeat (ir) technique. This approach coupled with a developed analytical platform allowed the floral phenotypic analysis of the circadian clock components.

The silencing of single circadian clock components had differential effects on the floral traits and elongation/flowering time of the transformed lines. Nevertheless, as in *Arabidopsis* there are not extensive works studying the effect on floral traits, this study in *N. attenuata* adds significant evidence on the differential regulation of single clock components.

The silencing of clock components had a strong effect on the opening time. The afore mentioned association of flower opening and TOC1 expression can be observed with irLHY line, where the opening time starts earlier at 14 h matching its peak expression of TOC1. Also the BA emission was set in advance at 16 h, correlating with the early corolla limb expansion, i.e. opening. This couldn't be studied through the irZTL clock-silenced lines in the glasshouse despite it had a slight earlier accumulation of TOC1 transcript in the corolla limb tissue. The reason is a partial flower opening; hence the BA emission cannot be correlated with the TOC1 transcript earlier accumulation as both factors cannot be separated. Instead, under field conditions irZTL flowers still emit low BA despite a fully open corolla; in this context, it was possible to separate opening from scent emission, so it is possible to suggest a control of ZTL through TOC1 over the BA emission.

The importance of TOC1 as a direct regulator of flower opening and BA emission is less supported by observing the irTOC1 flower phenotype, which doesn't present any strong deviation from the EV phenotype. Even so its absence causes a slight earlier opening, but more interestingly it extends the BA emission period by two extra hours in the night. The extended BA emission could be explained if TOC1 acts as a repressor of the genes involved in shutting down the emission. It is possible that other *PSEUDO-RESPONSE REGULATOR* genes participate in the BA regulation, possible with redundant functions of TOC1, like in the case of PRR7 and PRR9. These genes are proved to have partial redundancy in temperature responsiveness and are also regulated by ZTL (Salome & Mcclung 2005; Han 2006; Yamashino *et al.* 2008).

This type of functional redundancy is not observed in irZTL lines, which show an arrhythmic flower phenotype given that the opening is incomplete, BA is not emitted, and vertical movement is impeded. It is proven in *Arabidopsis* mutants, that LKP2 and FKF1 have partial redundancy over ZTL functions.

Even so, only a triple mutant of these three genes in *Arabidopsis* could produce a transcriptional arrhythmic plant in several circadian markers (Baudry *et al.* 2010). This evidence suggests that in *N. attenuata* several ZTL functions do not have a partial redundancy, at least in flowers, since the clock output is severely disrupted. This leads to propose that a ZTL protein interacts with many targets beside the well-known interaction with TOC1, PRRs genes and other genes. Further experiments should be conducted on proteomic arrays to test how broad ZTL interaction is in other metabolic pathways.

The comparative observations between glasshouse and field conditions showed that flower aperture is not only regulated by the circadian clock, but also through light signaling, because irZTL deficient flower opening was restored under the strong light intensity in the field plot at Utah. Additionally, the WT flower opening results under free running conditions showed a deficient closing compare to normal light/dark conditions. These two results allow to hypothesize that the flower aperture is partially circadian regulated because it requires light entrainment for tuning the correct clock cycle.

Elongation and flowering time consequences of silencing single clock components on glasshouse conditions were only observed in irTOC1 lines as depicted in Manuscript I, not on irZTL or irLHY lines. The results of these last lines do not match with the observed phenotypes of *A. thaliana* Columbia accession, like previously exemplified with Arabidopsis *toc1-1* mutant accession. This can be due to different genetic backgrounds of the Utah ecotype, the unknown gene redundancy, or a different clock regulation. It could be studied by crossing different clock deficient lines or using different ecotypes to observe the genetic importance of each component combination during the plant development towards reproductive stage. In perspective, it will be important to study and compare the plant development time by stages between glasshouse and field conditions because the strong entraining signals and abiotic stresses in the field make clock periods differences more visible, leading the plant to compromise its survival and reproduction, depending on the advantages of their clock.

Relevance of *Nicotiana attenuata* floral traits

The genus *Nicotiana* has developed a pollination interaction with the functional group of hawkmoths-hummingbirds since all of their described species have a tubular corolla with nectaries in the bottom that produces a nutritious reward. Either a hawkmoth or a hummingbird is equipped with a long collector organ to reach the bottom nectar, probiscides, and tongue, respectively. Color and size varies among *Nicotiana* species, as it is expected from an optimal foraging theory perspective because, despite the adaptations for a same functional group, the pollinator members vary in size and activity time and periods over latitudinal and altitudinal gradients that the *Nicotiana* species inhabit over three continents: America, Africa and Oceania(Goodspeed 1954).

Similar flower specializations are the scent emissions that several studies have measured, and then identified the major scents in certain *Nicotiana* species (Loughrin *et al.* 1991; Kolosova *et al.* 2001; Raguso *et al.* 2003). Unlike *N. attenuata,* other genus species major scent is benzyl alcohol, instead of benzyl acetone, but these species are also mainly pollinated by hawkmoths. This difference would require a hawkmoth approach to solve the attractiveness of both compounds and define the importance of the distinctiveness of BA.

Adaptation of vertical movement

Following Sttebin's principle of a trait preceding the appearance of an interaction, the mutations for vertical movement should have appeared first, but its circadian control is rather a question. Given the marked night-time outcrossing results between clock-silenced lines in the field, the circadian control over this trait is necessary for a successful reproduction. Since the vertical movement is not reported in other *Nicotiana* species, it can be assumed as a novel trait because the pedicel tissue initially developed uniformly, without producing any movement by a time regulation (Cortes Llorca *et al.* 2013).

Thus, a time separation of tissue growth will have lead to the movement, which poses the question if this was originally guided by the internal rhythms, or uncoupled and later controlled by the clock. The resemblance of the pedicel tissue transcripts for clock genes with the leaf tissue and the results from irZTL movement disruption will suggest that it originally started under clock control. An interesting question is if the growth difference that produces the movement started in its current pattern or inverted. It can be answered by comparing to the observations of *Nicotiana acuminata* movement in the glasshouse, which have an inverted vertical movement, not bending downwards but upwards. As we have two related species with inverted movement direction with a circadian pattern, it is more plausible that the original trait movement, phase, and direction appeared as it is currently observed. This is supported by the evidence that the clock silencing in *N. attenuata* doesn't invert the movement but shifts it or reduces it.

What can be argued is the amplitude and range of the vertical movement, as this might adapt overtime depending on the selective pressures of the habitat. Under this assumption, the amplitude can be a plastic trait, which adapts depending on the availability of pollinators that select those with stronger and marked different orientation between day and night. If in the original habitat both hawkmoths and hummingbirds were present, those that provide a better pollination service will have exerted a stronger selection, especially in the case of the hawkmoths that would have driven the amplitude selection towards an upward orientation at night. This would have exerted a negative pressure over those without a lower orientation during the day. The specimens with short amplitude would have allowed an easier recognition of the flower by the hummingbirds, but if their outcrossing output was lesser than the one mediated by hawkmoth, then it wouldn't have been selected in the Utah ecotype. In perspective, complementary studies should measure pollen removal and transfer to directly evaluate the pollinator efficiency.

Pollinators

The flowers of *N. attenuata* have a white color that is agreed by several studies to be attractive for hawkmoths (Sprayberry & Suver 2011). It is preferable for hawkmoths to fly under low light to avoid predation, e.g. bats, on which case the olfactive stimuli plays a strong role.

The emission of a floral bouquet in time plays a strong role in attracting *Manduca* hawkmoths to the flower sources. As it was observed in the field, the lack of scent emission ends up in reduced pollination service by *Manduca* spp., demonstrated by the lack of BA in irZTL flowers. The importance of a correct emission time is crosschecked with the irLHY flowers, where an earlier strong emission did not attract them. The presence or activity period could not be disentangled with the performed experiments, since it was not possible to know if Manduca hawkmoths were present at a given time in the plot, either active or inactive, as they could only be identified by visual contact when flying. An initial hypothesis of the high outcrossing results at glasshouse conditions with irLHY flowers was that in a contained space with low airflow the scent emission could induce an earlier visitation by *M. sexta*. Given the low outcrossing results at field conditions and the lack of any observed *M. sexta* at dusk, it can be concluded that an earlier scent emission does not stimulate the hawkmoths to start its activity period at earlier time, independently of proximity or activity period.

Since *N. attenuata* flowers reach their upward orientation before dusk, the original adaptation would have been for pollinators active from dusk onwards. If not, the chosen flowers by selection would have had a longer clock loop that synchronizes better with the currently more efficient pollinator at Utah, *Manduca sexta* hawkmoth. From the corolla opening point of view, this one reaches its full opening after the upward orientation is achieved, and even later the peak of BA emission.

At the field station in Utah, the hawkmoth *Hyles lineata* is also present. It acts as one pollinator of *N. obtusifolia*, visiting the flowers before and after dusk, time after which it is infrequent and can also act as *N. attenuata* pollinator(Kessler *et al.* 2013). This outcrossing viability with both pollinators suggests an old interaction with a single pollinator with an earlier and longer period of activity or a pollinator assemblage. At the present time, it has been revealed that several long considered specialized flowers depend quite much on pollinator assemblages (Waser *et al.* 1996). This could be the case of *N. attenuata* with regard to hawkmoths, because not only *Manduca* spp. will be preferred, but a wider range of hawkmoths with different activity periods prior dusk onwards. The attraction of most efficient pollinators could work through the floral traits sequence display (upward orientation->opening->scents) without stopping the attraction of other hawkmoths that can respond to traits separately. A plasticity to attract several hawkmoths and also graduate for the most efficient will be a good adaptation in a wider range of habitats with different hawkmoth assemblages. These remain to be tested, for example, by transplanting ecotypes and studying the pollinator community according to the floral traits.

Chapter 6 – Discussion

Hummingbirds are the main day-time pollinator considered in this study because it feeds on nectar, independently of the pollen presence. As previously demonstrated by Kessler *et al.* (Kessler *et al.* 2010), *Archilochus alexandri* is the principal day-time pollinator, which preferentially feeds on morning open flowers because the nectar is not consumed overnight by hawkmoths. In the field, an altered clock was favored by day-time pollinators, possibly for two causes: the lack of BA emission at night, which is a main attractant of hawkmoths, and the lack of vertical movement during day, which make flowers in an upward orientation more conspicuous.

The importance of BA has been tested in studies of Kessler (Kessler *et al.* 2008; Kessler & Baldwin 2011) using silenced lines deficient in BA production, where a lesser capsule production was found in comparison to WT phenotype flowers. This supports irZTL capsules production, as that was the only factor differentiating it from the EV and irLHY in the field at night-time. This silenced line shows how ZTL gene plays an important role in regulating BA emission and the disadvantageous reproductive effects of its disruption. By comparing to morning open flowers that have a partial corolla closure in the field after mid-morning, but bend downwards as any night opening flower, irZTL flowers kept an upward orientation that made them more conspicuous and attractive to hummingbird as morning flowers during the partial closure hours. Considering the fact that hummingbirds can learn overtime to choose the most nutritious source, it gives them the advantage of, once irZTL flowers are recognized, continuously visit them in the mornings to feed on its not removed nectar. In this way, the hummingbird foraging behavior facilitates a higher outcrossing during day-time, when normally no other plant with a functional clock will be beneficiated.

Nectar robbers and pollen feeders

In a more holistic view, the development and conservation of a flower vertical movement could be hypothesized to respond to multiple selective pressures beside the positive pressure of pollinators. In the 2012 and 2013 field seasons I could observe carpenter bees (*Xylopa* spp.) visiting *N. attenuata* flowers at dawn and dusk to rob the nectar by punctures at the flower base, which has been previously described in studies by Kessler(Kessler *et al.* 2008, 2010). In the same manner during day, several species commonly known as "sweat bees" visited open flowers searching for pollen to collect. If the vertical movement would have developed to avoid nectar robbers and pollen feeders by not advertising flowers in upward orientation, it would have required a longer day-time loop. The flower reaches its upward position at subjective dusk (18 h) and moves downward after dawn, not avoiding carpenter bees or sweat bees, rendering it as an ineffective measure. The movement will be wasteful in energy terms, as it is not a good trade-off between energy usage and avoidance. In the same sense, the flower aperture time should have had a stricter period, as its partial opening happens early enough to attract the robbers and feeders and, similarly, its partial closing happens too late to avoid them.

Given that both visual traits do not occur on a stricter time period, after dusk and prior to dawn, it does not avoid the time of carpenter bees and sweat bees foraging behavior. For this reason, it can be inferred that the fine circadian control of the floral traits developed to synchronize in response to the pollinators selection pressure, instead of the negative pressure that nectar robbers and pollen feeders would exert in the floral traits timing.

Conclusions

As any other organism regulated by a circadian clock, the solanaceous plant *N. attenuata* uses its endogenous clock to mark the day phases and prepare itself in advance, but unlike animals it cannot directly transfer its gametes. In this thesis it is demonstrated that *N. attenuata* also counts with an internal circadian clock (Manuscript I) that is conserved and works as suggested by the current accepted plant clock model. It uses its internal clock to regulate its floral traits (Manuscript II) and synchronizes them with the activity period of its night pollinator *M. sexta* for successful outcrossing (Manuscript III). Besides describing how the floral traits are regulated by the circadian clock, this thesis provides first evidence of the ecological relevance of the circadian control to mediate pollination using clock-silenced lines with dysfunctional clocks. Also, I demonstrated how an altered clock can be disadvantageous in a standard pollination interaction but advantageous under other context involving alternative pollination services. These new evidences of the circadian clock effects in flower and its synchronization importance to successful reproduction point to new paths of study in pollination ecology, and the direct mechanisms of clock control, with further ramifications to improve crop pollination services.

To conclude, pollinator-dependent plants confront a time challenge, over their natural history and day to day, of developing and synchronizing traits that will attract pollinators in exchange of a reward. The flower effort to attract pollinators leads it to finely tune the traits display with the pollinator activity period, shaping not only a morphological but also a time pollination syndrome.

Summary

Like most of all known organisms, the angiosperm plants need to keep a daily rhythm to synchronize their inner functions according to the availability of energy and nutrients; even due to its sessile nature it's especially important for a large group to find ways to reproduce by outcrossing in order to prevent deleterious selfing. The endogenous circadian clock regulates metabolic processes related to the synchronization of plant traits that allows its survival and successful reproduction in pollinator dependent plants.

In this work, the circadian clock genes in *N. attenuata* were identified and found to have a conserved function, as demonstrated by yeast two-hybrid assays and a similar hypocotyl length phenotypes on transformed *Arabidopsis thaliana* plants using *N. attenuata* transferred genes. Additionally, under long day conditions NaLHY and NaTOC1 peak at dawn and dusk, respectively, without NaZTL having marked oscillations just like in other plant clock models. The silencing of circadian genes in *N. attenuata* using inverted repeat (ir) technique showed similar clock transcription profile alterations of NaCAB2 and hypocotyl growth alteration as in the established clock model. Nevertheless, irTOC1 plants had a delayed elongation and flowering under long day conditions unlike some other model plant accessions, but this can be attributed to a different habitat and genetic background, agreeing with previous observations in cross latitudinal studies of other species.

Vertical movement, as observed in *N. attenuata*, appears as an independent trait in its genus, which temporally excludes other pollinators by reducing the flower conspicuousness. Floral traits are under circadian control and are relevant to synchronize the flower display and scent emission with its pollinator *Manduca sexta*. The correct synchronization improves outcrossing by *M. sexta* in the field since dysrhythmia by shifts of the floral traits resulted

Zusammenfassung

ineffective to improve the outcrossing as in irLHY case. However, a lack of regulation caused basically arrhythmic floral traits since irZTL had unexpected advantages during day-time outcrossing in the field due to the recruitment of day-active pollinators that provided similar reproductive results as night-pollinator dependent plants.

Zusammenfassung

Wie die meisten aller bekannten Organismen müssen bedecktsamige Pflanzen (Angiospermen) einen Tagesrhythmus einhalten um ihre inneren Funktionen mit der Verfügbarkeit von Licht und Nährstoffen zu synchronisieren. Aufgrund der sessilen Lebensweise ist es besonders für Pflanzen wichtig Wege für eine Auskreuzung zu finden um eine Selbstbestäubung zu verhindern. Die innere circadiane Uhr regelt Stoffwechselvorgänge und anderen Merkmale, die ein Überleben und eine erfolgreiche Fortpflanzung für Pflanzen ermöglicht, welche auf Bestäubung durch Insekten angewiesen sind.

In dieser Arbeit wurden Gene der circadianen Uhr von *N. attenuata* identifiziert und kloniert und eine konservierte Funktion konnte mittels eines Hefe-Zwei-Hybrid-Systems und der Expression in *Arabidopsis thaliana*-Pflanzen nachgewiesen werden, welche mit den von *N. attenuata* übertragenen Genen einen ähnliche Phänotyp in der Hypokotyllänge aufwiesen. Zusätzlich zeigten die Gene NaLHY und NaTOC1 unter Langtagsbedingungen ihr Expressionsmaximum jeweils bei Dämmerung oder Morgengrauen, ohne dass jedoch NaZTL eine Oszillation zeigte, wie bereits in anderen Pflanzenmodellen gezeigt wurde. Das silencen der circadianen Gene in *N. attenuata* mittels inverted repeat (IR) Technik resultierte in einer ähnlichen Änderung des Transkriptionsprofils für NaCAB2 und dem Hypokotyl Wachstum wie auch in anderen Modellsystemen gezeigt wurde. Jedoch zeigten irTOC1 Pflanzen unter Langtagsbedingungen ein verzögertes Wachstum und Blütenbildung, welches im Gegensatz zu den Ergebnissen von einigen anderen Modellpflanzen steht. Dies kann jedoch den unterschiedlichen Lebensräumen und dem genetischen Hintergrund zugeschrieben werden, wie bereits früher zwischen verschiedenen Spezies beobachtet werden konnte.

Zusammenfassung

Eine vertikale Bewegung der Blüten, wie in *N. attenuata* beobachtet, scheint ein spezielles Merkmal dieser Pflanzengattung zu sein, die es ermöglicht bestimmte Bestäuber durch eine Reduzierung der Blütenpräsenz auszuschließen. Verschiedene Blütenmerkmale sind unter circadianer Kontrolle und wichtig um den Duft und die Präsenz der Blüte mit der Anwesenheit des Bestäubers (*Manduca sexta*) zu synchronisieren. In einem Feldversuch konnte gezeigt werden dass eine funktionierende Synchronisation die Auskreuzungsrate durch *M. sexta* erhöhen kann, wohingegen die asynchronen Blüteneigenschaften wie bei irLHY der Fall, die Auskreuzungsrate verringern kann. Allerdings zeigte sich auch, dass ein Mangel an Regulierung und grundsätzlich arrhythmische Blüteneigenschaften wie bei irZTL der Fall, auch unerwartete Vorteile bringen kann, da durch die Rekrutierung von tagaktiven Bestäubern, welche ähnlich effizient waren wie nachtaktive Bestäuber, ein Auskreuzen gewährleistet wurde.

Bibliography

Adams, S. & Carré, I.A. (2011). Downstream of the plant circadian clock: output pathways for the control of physiology and development. *Essays in Biochemistry*, 49, 53–69.

Aizen, M.A. (2003). Down-Facing Flowers, Hummingbirds and Rain. *Taxon*, 52, 675.

Aizen, M.A., Morales, C.L. & Morales, J.M. (2008). Invasive mutualists erode native pollination webs. *PLoS biology*, 6, e31.

Alabadi, D., Oyama, T., Yanovsky, M.J., Harmon, F.G., Mas, P. & Kay, S.A. (2001). Reciprocal regulation between TOC1 and LHY/CCA1 within the *Arabidopsis circadian* clock. *Science*, 293, 880–883.

Albrecht, U. & Eichele, G. (2003). The mammalian circadian clock. *Current opinion in genetics & development*, 13, 271–7.

Ashman, T.-L. & Schoen, D.J. (1994). How long should flowers live? *Nature*, 371.

Baldwin, I.T., Staszakkozinski, L. & Davidson, R. (1994). Up in smoke. 1. Smoke-derived germination cues for postfire annual, *Nicotiana-attenuata* Torr Ex Watson. *Journal of Chemical Ecology*, 2345 – 2371.

Barak, S., Tobin, E.M., Green, R.M., Andronis, C. & Sugano, S. (2000). All in good time: the *Arabidopsis* circadian clock. *Trends Plant Sci*, 5, 517–522.

Baudry, A., Ito, S., Song, Y.H., Strait, A.A., Kiba, T., Lu, S., et al. (2010). F-box proteins FKF1 and LKP2 act in concert with ZEITLUPE to control *Arabidopsis* clock progression. *Plant Cell*, 22, 606–622.

Bhattacharya, S. & Baldwin, I.T. (2012). The post-pollination ethylene burst and the continuation of floral advertisement are harbingers of non-random mate selection in *Nicotiana attenuata*. *The Plant Journal*, 71, 587–601.

Buchmann, S.L. (1987). The Ecology of Oil Flowers and their Bees. *Annual Review of Ecology and Systematics*, 18, 343–369.

Bunning, E. (1956). Endogenous rhythms in plants. *Annual Review of Plant Physiology*, 7, 71–90.

Clarke, D., Whitney, H., Sutton, G. & Robert, D. (2013). Detection and learning of floral electric fields by bumblebees. *Science*, 340, 66–9.

Climate 101. (2013). Climate 101 [WWW Document]. URL http://www.climate101.org/2013/03/the-great-basin/.

Corellou, F., Schwartz, C., Motta, J.-P., Djouani-Tahri, E.B., Sanchez, F. & Bouget, F.-Y. (2009). Clocks in the green lineage: comparative functional analysis of the circadian architecture of the *Picoeukaryote Ostreococcus*. *The Plant Cell*, 21, 3436–3449.

Cortes Llorca, L., Yon, F., Rothe, E., Kim, S.-G. & Baldwin, I.T. (2013). Circadian movement of flowers via biased auxin flow in pedicel of *Nicotiana attenuata*. (Unpublished)

Covington, M.F., Panda, S., Liu, X.L., Strayer, C. a, Wagner, D.R. & Kay, S. a. (2001). ELF3 modulates resetting of the circadian clock in *Arabidopsis*. *The Plant cell*, 13, 1305–15.

Dinh, S.T., Gális, I. & Baldwin, I.T. (2013). UVB radiation and 17-hydroxygeranyllinalool diterpene glycosides provide durable resistance against mirid (*Tupiocoris notatus*) attack in field-grown *Nicotiana attenuata* plants. *Plant, cell & environment*, 36, 590–606.

Dodd, A.N., Salathia, N., Hall, A., Kevei, E., Toth, R., Nagy, F., et al. (2005). Plant circadian clocks increase photosynthesis, growth, survival, and competitive advantage. *Science*, 309, 630–633.

Doherty, C.J. & Kay, S. a. (2010). Circadian control of global gene expression patterns. *Annual review of genetics*, 44, 419–44.

Van Doorn, W.G. & Van Meeteren, U. (2003). Flower opening and closure: a review. *Journal of Experimental Botany*, 54, 1801–1812.

Dunlap, J.C. (1996). Genetic and molecular analysis of circadian rhythms. *Annu Rev Genet*, 30, 579–601.

Edgar, R.S., Green, E.W., Zhao, Y., van Ooijen, G., Olmedo, M., Qin, X., et al. (2012). Peroxiredoxins are conserved markers of circadian rhythms. *Nature*, 485, 459–464.

Farre, E.M., Harmer, S.L., Harmon, F.G., Yanovsky, M.J. & Kay, S.A. (2005). Overlapping and distinct roles of PRR7 and PRR9 in the *Arabidopsis* circadian clock. *Curr Biol*, 15, 47–54.

Fenster, C.B., Armbruster, W.S., Wilson, P., Dudash, M.R. & Thomson, J.D. (2004). Pollination Syndromes and Floral Specialization. *Annual Review of Ecology, Evolution, and Systematics*, 35, 375–403.

Filichkin, S.A., Breton, G., Priest, H.D., Dharmawardhana, P., Jaiswal, P., Fox, S.E., et al. (2011). Global profiling of rice and poplar transcriptomes highlights key conserved circadian-controlled pathways and cis-regulatory modules. *PLoS One*, 6, e16907.

Fournier-Level, a, Korte, A., Cooper, M.D., Nordborg, M., Schmitt, J. & Wilczek, a M. (2011). A map of local adaptation in *Arabidopsis thaliana*. *Science*, 334, 86–9.

Fründ, J., Dormann, C.F. & Tscharntke, T. (2011a). Linné's floral clock is slow without pollinators flower closure and plant-pollinator interaction webs. *Ecology letters*, 14, 896–904.

Fründ, J., Dormann, C.F. & Tscharntke, T. (2011b). Linné's floral clock is slow without pollinators flower closure and plant-pollinator interaction webs. *Ecology Letter*, 14, 896–904.

Fulton, M. & Hodges, S.A. (1999). Floral isolation between *Aquilegia formosa* and *Aquilegia pubescens*. *Proceedings of the Royal Society B-Biological Sciences*, 266.

Gase, K., Weinhold, A., Bozorov, T., Schuck, S. & Baldwin, I.T. (2011). Efficient screening of transgenic plant lines for ecological research. *Molecular ecology resources*, 11, 890–902.

Gendron, J.M., Pruneda-Paz, J.L., Doherty, C.J., Gross, A.M., Kang, S.E. & Kay, S.A. (2012). *Arabidopsis* circadian clock protein, TOC1, is a DNA-binding transcription factor. *Proceedings of the Natural Academy of Science USA*, 109, 3167–3172.

Golden, S.S. & Canales, S.R. (2003). Cyanobacterial circadian clocks - timing is everything. *Nat Rev Microbiol*, 1, 191–199.

Goodspeed, T.H. (1954). The genus Nicotiana: origins, relationships, and evolution of its species in the light of their distribution, morphology, and cytogenetics. Chronica Botanica Co.

Goodspeed, D., Chehab, E.W., Min-Venditti, A., Braam, J. & Covington, M.F. (2012). *Arabidopsis* synchronizes jasmonate-mediated defense with insect circadian behavior. *Proceedings of the Natural Academy of Science*, 109, 4674–4677.

Goyret, J. (2010). Look and touch: multimodal sensory control of flower inspection movements in the nocturnal hawkmoth *Manduca sexta*. *The Journal of experimental biology*, 213, 3676–82.

Grant, V. (1949). Pollination Systems as Isolating Mechanisms in Angiosperms. *Evolution*, 3, 82–97.

Grant, V. (1952). Isolation and Hybridization Between *Aquilegia Formosa* and *A. Pubescens*. In: *The Origin of Adaptations*. Columbia University Press.

Green, R.M., Tingay, S., Wang, Z.Y. & Tobin, E.M. (2002). Circadian rhythms confer a higher level of fitness to *Arabidopsis* plants. *Plant Physiol*, 129, 576–584.

Han, L. (2006). *F-BOX PROTEIN , ZEITLUPE , IN THE ARABIDOPSIS CIRCADIAN CLOCK. Genetic Analysis*. Thesis. The Ohio University

Hancock, A.M., Brachi, B., Faure, N., Horton, M.W., Jarymowycz, L.B., Sperone, F.G., et al. (2011). Adaptation to climate across the *Arabidopsis thaliana* genome. *Science*, 334, 83–6.

Harder, L.D. & Barrett, S.C.H. (2006). *Ecology and Evolution of Flowers*. Oxford University Press, New York.

Harmer, S.L. (2009). The circadian system in higher plants. *Annual review of plant biology*, 60, 357–77.

Helfer, A., Nusinow, D.A., Chow, B.Y., Gehrke, A.R., Bulyk, M.L. & Kay, S.A. (2011). LUX ARRHYTHMO encodes a nighttime repressor of circadian gene expression in the *Arabidopsis* core clock. *Current biology : CB*, 21, 126–33.

Hoballah, M.E., Stuurman, J., Turlings, T.C.J., Guerin, P.M., Connétable, S. & Kuhlemeier, C. (2005). The composition and timing of flower odour emission by wild *Petunia axillaris* coincide with the antennal perception and nocturnal activity of the pollinator *Manduca sexta*. *Planta*, 222, 141–50.

Hodges, S.A. & Arnold, M.L. (1994). Columbines: a geographically widespread species flock. *Proceedings of the National Academy of Sciences USA*, 91, 5129–5132.

Hodges, S.A. & Arnold, M.L. (1995). Spurring Plant Diversification: Are Floral Nectar Spurs a Key Innovation? *Proceedings of the Royal Society B-Biological Sciences*, 262, 343–348.

Hodges, S.A., Fulton, M., Yang, J.Y. & Whittall, J.B. (2004). Verne Grant and evolutionary studies of *Aquilegia*. *New Phytologist*, 161, 113–120.

Hsu, P.Y., Devisetty, U.K. & Harmer, S.L. (2013). Accurate timekeeping is controlled by a cycling activator in *Arabidopsis*. *eLife Sciences*, 2, e00473.

Huang, W., Pérez-García, P., Pokhilko, A., Millar, A.J., Antoshechkin, I., Riechmann, J.L., et al. (2012). Mapping the core of the Arabidopsis circadian clock defines the network structure of the oscillator. *Science*, 336, 75–79.

Hurlbert, A.H., Hosoi, S.A., Temeles, E.J. & Ewald, P.W. (1996). Mobility of *Impatiens capensis* flowers: effect on pollen deposition and hummingbird foraging. *Oecologia*, 105, 243–246.

Hut, R.A., Paolucci, S., Dor, R., Kyriacou, C.P. & Daan, S. (2013). Latitudinal clines: an evolutionary view on biological rhythms. *Proceedings of the Royal Society B-Biological Sciences*.

Imaizumi, T., Schultz, T.F., Harmon, F.G., Ho, L.A. & Kay, S.A. (2005). FKF1 F-box protein mediates cyclic degradation of a repressor of CONSTANS in *Arabidopsis*. *Science*, 309, 293–297.

Itoh, H. & Izawa, T. (2013). The Coincidence of Critical Day Length Recognition for Florigen Gene Expression and Floral Transition under Long-Day Conditions in Rice. *Molecular plant*, 6, 635–49.

Izawa, T. (2012). Physiological significance of the plant circadian clock in natural field conditions. *Plant, cell & environment*, 35, 1729–41.

Izawa, T., Oikawa, T., Sugiyama, N., Tanisaka, T., Yano, M. & Shimamoto, K. (2002). Phytochrome mediates the external light signal to repress FT orthologs in photoperiodic flowering of rice. *Genes Development*, 16, 2006–2020.

James, A.B., Monreal, J.A., Nimmo, G.A., Kelly, C.L., Herzyk, P., Jenkins, G.I., et al. (2008). The circadian clock in *Arabidopsis* roots is a simplified slave version of the clock in shoots. *Science*, 322, 1832–1835.

Jones, C.E. & Little, R.J. (1983). *Handbook of experimental pollination biology*. Van Nostrand Reinhold, New York.

Kaldis, A.-D., Kousidis, P., Kesanopoulos, K. & Prombona, A. (2003). Light and circadian regulation in the expression of LHY and lhcb genes in *Phaseolus vulgaris*. *Plant Molecular Biology*, 52, 981–997.

Kaldis, A.-D. & Prombona, A. (2006). Synergy between the light-induced acute response and the circadian cycle: a new mechanism for the synchronization of the *Phaseolus vulgaris* clock to light. *Plant Molecular Biology*, 61, 883–95.

Kessler, D. (2012). Context dependency of nectar reward-guided oviposition. *Entomologia Experimentalis et Applicata*, 144, 112–122.

Kessler, D. & Baldwin, I.T. (2007). Making sense of nectar scents: the effects of nectar secondary metabolites on floral visitors of *Nicotiana attenuata*. *The Plant Journal*, 49, 840–854.

Kessler, D. & Baldwin, I.T. (2011). Back to the past for pollination biology. *Current opinion in plant biology*, 14, 429–34.

Kessler, D., Diezel, C. & Baldwin, I.T. (2010). Changing pollinators as a means of escaping herbivores. *Current Biology*, 20, 237–42.

Kessler, D., Gase, K. & Baldwin, I.T. (2008). Field experiments with transformed plants reveal the sense of floral scents. *Science*, 321, 1200–2.

Kessler, D., Yon, F., Schuman, M., Joo, Y., Kallenbach, M., Diezel, C., et al. (2013). Pollination and oviposition trials in the Isserstedt tent. Institute Symposium at Max-Planck-Institute for Chemical Ecology

Khan, S., Rowe, S. & Harmon, F. (2010). Coordination of the maize transcriptome by a conserved circadian clock. *BMC Plant Biol*, 10, 126.

Kiba, T., Henriques, R., Sakakibara, H. & Chua, N.-H. (2007). Targeted degradation of PSEUDO-RESPONSE REGULATOR5 by an SCFZTL complex regulates clock function and photomorphogenesis in *Arabidopsis thaliana*. *The Plant Cell*, 19, 2516–2530.

Kikis, E.A., Khanna, R. & Quail, P.H. (2005). ELF4 is a phytochrome-regulated component of a negative-feedback loop involving the central oscillator components CCA1 and LHY. *The Plant journal : for cell and molecular biology*, 44, 300–13.

Kim, S.G., Yon, F., Gaquerel, E., Gulati, J. & Baldwin, I.T. (2011). Tissue specific diurnal rhythms of metabolites and their regulation during herbivore attack in a native tobacco. *Nicotiana attenuata. PLoS One*, 6, e26214.

Kim, W.-Y., Fujiwara, S., Suh, S.-S., Kim, J., Kim, Y., Han, L., et al. (2007). ZEITLUPE is a circadian photoreceptor stabilized by GIGANTEA in blue light. *Nature*, 449, 356–60.

Kolosova, N. (2001). Regulation of Circadian Methyl Benzoate Emission in Diurnally and Nocturnally Emitting Plants. *THE PLANT CELL ONLINE*, 13, 2333–2347.

Krugel, T., Lim, M., Gase, K., Halitschke, R. & Baldwin, I.T. (2002). *Agrobacterium* -mediated transformation of Nicotiana attenuata, a model ecological expression system. *Chemoecology*, 12, 177–183.

Linnaeus, C. (1751). *Philosophia Botanica*. 1st edn. Stockholm & Amsterdam.

Liu, H., Wang, H., Gao, P., Xü, J., Xü, T., Wang, J., et al. (2009). Analysis of clock gene homologs using unifoliolates as target organs in soybean (*Glycine max*). *Journal of Plant Physiology*, 166, 278–89.

Lombardi, L.M. & Brody, S. (2005). Circadian rhythms in *Neurospora crassa*: clock gene homologues in fungi. *Fungal Genetic Biology*, 42, 887–892.

Loros, J.J. & Dunlap, J.C. (2001). Genetic and molecular analysis of circadian rhythms in N. eurospora. *Annual Review Physiology*, 63, 757–794.

Lou, P., Wu, J., Cheng, F., Cressman, L.G., Wang, X. & McClung, C.R. (2012). Preferential retention of circadian clock genes during diploidization following whole genome triplication in Brassica rapa. *The Plant Cell*, 24, 2415–2426.

Loughrin, J.H., Hamilton-Kemp, T.R., Andersen, R.A. & Hildebrand, D.F. (1991). Circadian rhythm of volatile emission from flowers of *Nicotiana sylvestris* and *N. suaveolens*. *Physiologia Plantarum*, 83, 492–496.

Más, P., Alabadí, D., Yanovsky, M.J., Oyama, T. & Kay, S.A. (2003). Dual role of TOC1 in the control of circadian and photomorphogenic responses in *Arabidopsis*. *The Plant Cell*, 15, 223–236.

Mas, P., Kim, W.-Y., Somers, D.E. & Kay, S.A. (2003). Targeted degradation of TOC1 by ZTL modulates circadian function in *Arabidopsis thaliana*. *Nature*, 426, 567–570.

Matsushika, A., Makino, S., Kojima, M. & Mizuno, T. (2000). Circadian waves of expression of the APRR1/TOC1 family of pseudo-response regulators in *Arabidopsis thaliana*: insight into the plant circadian clock. *Plant Cell Physiology*, 41, 1002–1012.

McClung, C.R. (2006). Plant circadian rhythms. *The Plant Cell*, 18, 792–803.

McClung, C.R. (2010). A modern circadian clock in the common angiosperm ancestor of monocots and eudicots. *BMC Biology*, 8, 55.

McClung, C.R. (2013). Beyond *Arabidopsis*: The circadian clock in non-model plant species. *Seminars in Cell & Developmental Biology*, 24, 430–436.

Michael, T.P., Salome, P.A., Yu, H.J., Spencer, T.R., Sharp, E.L., McPeek, M.A., et al. (2003). Enhanced fitness conferred by naturally occurring variation in the circadian clock. *Science*, 302, 1049.

Millar, A.J., Carre, I.A., Strayer, C.A., Chua, N.H. & Kay, S.A. (1995). Circadian clock mutants in arabidopsis identified by luciferase imaging. *Science*, 267, 1161.

Miwa, K., Serikawa, M., Suzuki, S., Kondo, T. & Oyama, T. (2006). Conserved expression profiles of circadian clock-related genes in two lemna species showing long-day and short-day photoperiodic flowering responses. *Plant Cell Physiology*, 47, 601–612.

Mizoguchi, T., Wheatley, K., Hanzawa, Y., Wright, L., Mizoguchi, M., Song, H.-R., et al. (2002). LHY and CCA1 are partially redundant genes required to maintain circadian rhythms in *Arabidopsis*. *Developmental Cell*, 2, 629–641.

Murakami, M., Tago, Y., Yamashino, T. & Mizuno, T. (2007). Comparative overviews of clock-associated genes of *Arabidopsis thaliana* and *Oryza sativa*. *Plant & cell physiology*, 48, 110–21.

Nagano, A.J., Sato, Y., Mihara, M., Antonio, B.A., Motoyama, R., Itoh, H., et al. (2012). Deciphering and Prediction of Transcriptome Dynamics under Fluctuating Field Conditions. *Cell*, 151, 1358–1369.

Nagel, D.H. & Kay, S.A. (2012). Complexity in the wiring and regulation of plant circadian networks. *Current Biology*, 22, R648–R657.

Nelson, D.C., Lasswell, J., Rogg, L.E., Cohen, M.A. & Bartel, B. (2000). FKF1, a clock-controlled gene that regulates the transition to flowering in *Arabidopsis*. *Cell*, 101, 331–340.

Nitta, K., Yasumoto, A.A. & Yahara, T. (2010). Variation of flower opening and closing times in F1 and F2 hybrids of daylily (*Hemerocallis fulva*; Hemerocallidaceae) and nightlily (*H. Citrina*). *American Journal of Botany*, 97, 261–267.

Niwa, Y., Ito, S., Nakamichi, N., Mizoguchi, T., Niinuma, K., Yamashino, T., *et al.* (2007). Genetic linkages of the circadian clock-associated genes, TOC1, CCA1 and LHY, in the photoperiodic control of flowering time in *Arabidopsis thaliana*. *Plant & cell physiology*, 48, 925–37.

Niwa, Y., Yamashino, T. & Mizuno, T. (2009). The circadian clock regulates the photoperiodic response of hypocotyl elongation through a coincidence mechanism in *Arabidopsis thaliana*. *Plant and Cell Physiology*, 50, 838.

Nusinow, D.A., Helfer, A., Hamilton, E.E., King, J.J., Imaizumi, T., Schultz, T.F., *et al.* (2011). The ELF4-ELF3-LUX complex links the circadian clock to diurnal control of hypocotyl growth. *Nature*, 475, 398–402.

O'Neill, J.S. & Reddy, A.B. (2011). Circadian clocks in human red blood cells. *Nature*, 469, 498–503.

Okada, R., Kondo, S., Satbhai, S.B., Yamaguchi, N., Tsukuda, M. & Aoki, S. (2009). Functional characterization of CCA1/LHY homolog genes, PpCCA1a and PpCCA1b, in the moss *Physcomitrella patens*. *The Plant journal : for cell and molecular biology*, 60, 551–63.

Overland, L. (1960). Endogenous rhythm in opening and odor of flowers of Cestrum nocturnum. *American Journal of Botany*, 47, 378–382.

Panda, S., Antoch, M.P., Miller, B.H., Su, A.I., Schook, A.B., Straume, M , *et al.* (2002). Coordinated Transcription of Key Pathways in the Mouse by the Circadian Clock. *Cell*, 109, 307–320.

Pokhilko, A., Fernández, A.P., Edwards, K.D., Southern, M.M., Halliday, K.J. & Millar, A.J. (2012). The clock gene circuit in *Arabidopsis* includes a repressilator with additional feedback loops. *Molecular systems biology*, 8, 574.

Pokhilko, A., Mas, P. & Millar, A.J. (2013). Modelling the widespread effects of TOC1 signalling on the plant circadian clock and its outputs. *BMC Systems Biology*, 7, 23.

Potts, S.G., Biesmeijer, J.C., Kremen, C., Neumann, P., Schweiger, O. & Kunin, W.E. (2010). Global pollinator declines: trends, impacts and drivers. *Trends in Ecology & Evolution*, 25, 345–353.

Raguso, R. a, Levin, R. a, Foose, S.E., Holmberg, M.W. & McDade, L. a. (2003). Fragrance chemistry, nocturnal rhythms and pollination "syndromes" in *Nicotiana*. *Phytochemistry*, 63, 265–284.

Raguso, R.A. (2004). Flowers as sensory billboards: progress towards an integrated understanding of floral advertisement. *Current opinion in Plant Biology*, 7, 434–40.

Ramos, A., Pérez-Solís, E., Ibáñez, C., Casado, R., Collada, C., Gómez, L., *et al.* (2005). Winter disruption of the circadian clock in chestnut. *Proceedings of the National Academy of Sciences USA*, 102, 7037–42.

Resco, V., Hartwell, J. & Hall, A. (2009). Ecological implications of plants' ability to tell the time. *Ecology Letters*, 12, 583–592.

Roda, A., Halitschke, R., Steppuhn, A. & Baldwin, I.T. (2004). Individual variability in herbivore-specific elicitors from the plant's perspective. *Molecular Ecology*, 13, 2421–33.

Roden, L.C. & Ingle, R.A. (2009). Lights, rhythms, infection: the role of light and the circadian clock in determining the outcome of plant-pathogen interactions. *The Plant Cell*, 21, 2546–52.

Roenneberg, T. (1994). The Gonyaulax circadian system: Evidence for two input pathways and two oscillators. In: *Evolution of Circadian Clock* (eds. Hiroshige, T. & Honma, K.-I.). Hokkaido University Press, Sapporo, pp. 3–20.

Roenneberg, T. & Mittag, M. (1996). The circadian program of algae. *Seminars in Cell & Developmental Biology*, 7, 753–763.

Salome, P.A. & McClung, C.R. (2005). PSEUDO-RESPONSE REGULATOR 7 and 9 are partially redundant genes essential for the temperature responsiveness of the *Arabidopsis* circadian clock. *Plant Cell*, 17, 791–803.

Sanchez, A., Shin, J. & Davis, S.J. (2011). Abiotic stress and the plant circadian clock. *Plant Signaling & Behavior*, 6, 223–231.

Sawa, M., Nusinow, D.A., Kay, S.A. & Imaizumi, T. (2007). FKF1 and GIGANTEA complex formation is required for day-length measurement in *Arabidopsis*. *Science*, 318, 261–265.

Schaffer, R., Ramsay, N., Samach, A., Corden, S., Putterill, J., Carre, I.A., *et al.* (1998). The late elongated hypocotyl mutation of *Arabidopsis* disrupts circadian rhythms and the photoperiodic control of flowering. *Cell*, 93, 1219–1229.

Schäfer, M., Fischer, C., Baldwin, I.T. & Meldau, S. (2011). Grasshopper oral secretions increase salicylic acid and abscisic acid levels in wounded leaves of *Arabidopsis thaliana*. *Plant Signaling & Behavior*, 6, 1256–8.

Seo, P.J., Park, M.-J., Lim, M.-H., Kim, S.-G., Lee, M., Baldwin, I.T., et al. (2012). A self-regulatory circuit of CIRCADIAN CLOCK-ASSOCIATED1 underlies the circadian clock regulation of temperature responses in Arabidopsis. *The Plant Cell*.

Somers, D.E., Webb, a a, Pearson, M. & Kay, S. a. (1998). The short-period mutant, toc1-1, alters circadian clock regulation of multiple outputs throughout development in *Arabidopsis thaliana*. *Development*, 125, 485–94.

Somers, D.E. (1999). The Physiology and Molecular Bases of the Plant Circadian Clock. *Plant Physiology*, 121, 9–20.

Somers, D.E., Kim, W.Y. & Geng, R. (2004a). The F-box protein ZEITLUPE confers dosage-dependent control on the circadian clock, photomorphogenesis, and flowering time. The *Plant Cell*, 16, 769–782.

Somers, D.E., Schultz, T.F., Milnamow, M. & Kay, S.A. (2000). ZEITLUPE Encodes a Novel Clock-Associated PAS Protein from *Arabidopsis*. *Cell*, 101, 319–329.

Somers, D.E., Webb, A., Pearson, M. & Kay, S.A. (1998). The short-period mutant, toc1-1, alters circadian clock regulation of multiple outputs throughout development in *Arabidopsis thaliana*. *Development*, 125, 485–494.

Sprayberry, J.D.H. & Daniel, T.L. (2007) Flower tracking in hawkmoths: behavior and energetics. *The Journal of experimental biology*, 210, 37–45.

Sprayberry, J.D.H. & Suver, M. (2011). Hawkmoths' innate flower preferences: a potential selective force on floral biomechanics. *Arthropod-Plant Interactions*, 263–268.

Staiger, D., Shin, J., Johansson, M. & Davis, S. (2013). The circadian clock goes genomic. *Genome Biology*, 14, 208.

Stebbins, G.L. (1970). Adaptive Radiation of Reproductive Characteristics in Angiosperms, I: Pollination Mechanisms. *Annual Review of Ecology and Systematics*, 1, 307–326.

Steppuhn, A., Schuman, M.C. & Baldwin, I.T. (2008). Silencing jasmonate signalling and jasmonate-mediated defences reveals different survival strategies between two *Nicotiana attenuata* accessions. *Molecular Ecology*, 17, 3717–3732.

Stitt, M. & Zeeman, S.C. (2012). Starch turnover: pathways, regulation and role in growth. *Current opinion in plant biology*, 15, 282–92.

Strayer, C., Oyama, T., Schultz, T.F., Raman, R., Somers, D.E., Más, P., et al. (2000). Cloning of the *Arabidopsis* clock gene TOC1, an autoregulatory response regulator homolog. *Science*, 289, 768.

Sweeney, B.M. (1963). Biological clocks in plants. *Annual Review of Plant Physiology*, 14, 411–440.

Tadey, M. & Aizen, M.A. (2001). Why do flowers of a hummingbird-pollinated mistletoe face down? *Functional Ecology*, 15, 782–790.

Takata, N., Saito, S., Saito, C.T., Nanjo, T., Shinohara, K. & Uemura, M. (2009). Molecular phylogeny and expression of poplar circadian clock genes, LHY1 and LHY2. *The New Phytologist*, 181, 808–19.

Taylor, A., Massiah, A.J. & Thomas, B. (2010). Conservation of *Arabidopsis thaliana* photoperiodic flowering time genes in onion (*Allium cepa* L.). *Plant Cell & Physiology*, 51, 1638–1647.

Thain, S.C., Murtas, G., Lynn, J.R., McGrath, R.B. & Millar, A.J. (2002). The circadian clock that controls gene expression in *Arabidopsis* is tissue specific. *Plant Physiology*, 130, 102–10.

Ueda, H.R. (2006). Systems biology flowering in the plant clock field. *Molecular System Biology*, 2, 60.

Ushimaru, A. & Hyodo, F. (2005). Why do bilaterally symmetrical flowers orient vertically? Flower orientation influences pollinator landing behaviour. *Evolutionary Ecology Research*, 7, 151–160.

Voelckel, C., Schittko, U. & Baldwin, I.T. (2001). Herbivore-induced ethylene burst reduces fitness costs of jasmonate- and oral secretion-induced defenses in *Nicotiana attenuata*. *October*, 274–280.

Wang, W., Barnaby, J.Y., Tada, Y., Li, H., Tor, M., Caldelari, D., et al. (2011). Timing of plant immune responses by a central circadian regulator. *Nature*, 470, 110–114.

Wang, X., Wu, L., Zhang, S., Wu, L., Ku, L., Wei, X., et al. (2011c). Robust expression and association of ZmCCA1 with circadian rhythms in maize. *Plant Cell reports*, 30, 1261–72.

Wang, Z.-Y. & Tobin, E.M. (1998). Constitutive expression of the CIRCADIAN CLOCK ASSOCIATED 1 (CCA1) gene disrupts circadian rhythms and suppresses its own expression. *Cell*, 93, 1207–1217.

Waser, N.M., Chittka, L., Price, M. V, Williams, N.M. & Ollerton, J. (1996). Generalization in pollination systems, and why it matters. *Ecology*, 77, 1043–1060.

Wenden, B., Toner, D.L.K., Hodge, S.K., Grima, R. & Millar, a J. (2012). Spontaneous spatiotemporal waves of gene expression from biological clocks in the leaf. *Proceedings of the National Academy of Sciences USA*, 109, 6757–6762.

Williams, J.A. & Sehgal, A. (2001). Molecular components of the circadian system in drosophila. *Annual Review Physiology*, 63, 729–755.

Wu, J., Hettenhausen, C., Schuman, M.C. & Baldwin, I.T. (2008). A comparison of two *Nicotiana attenuata* accessions reveals large differences in signaling induced by oral secretions of the specialist herbivore *Manduca sexta*. *Plant Physiology*, 146, 927–39.

Wu, J., Kang, J.-H., Hettenhausen, C. & Baldwin, I.T. (2007). Nonsense-mediated mRNA decay (NMD) silences the accumulation of aberrant trypsin proteinase inhibitor mRNA in *Nicotiana attenuata*. *The Plant journal : for cell and molecular biology*, 51, 693–706.

Wu, J. & Baldwin, I.T. (2010). New insights into plant responses to the attack from insect herbivores. *Annual Review Genetics*, 44, 1–24.

Xu, X., Xie, Q. & McClung, C.R. (2010a). Robust circadian rhythms of gene expression in *Brassica rapa* tissue culture. *Plant Physiology*, 153, 841–850.

Xue, Z.G., Zhang, X.M., Lei, C.F., Chen, X.J. & Fu, Y.F. (2011). Molecular cloning and functional analysis of one ZEITLUPE homolog GmZTL3 in soybean. *Molecular Biology Reports*, 39, 1411–1418.

Yakir, E., Hassidim, M., Melamed-Book, N., Hilman, D., Kron, I. & Green, R.M. (2011). Cell autonomous and cell-type specific circadian rhythms in *Arabidopsis*. *The Plant Journal*, 68, 520–31.

Yakir, E., Hilman, D., Harir, Y. & Green, R.M. (2007). Regulation of output from the plant circadian clock. *The FEBS journal*, 274, 335–45.

Yamashino, T., Ito, S., Niwa, Y., Kunihiro, A., Nakamichi, N. & Mizuno, T. (2008). Involvement of *Arabidopsis* clock-associated pseudo-response regulators in diurnal oscillations of gene expression in the presence of environmental time cues. *Plant & Cell Physiology*, 49, 1839–50.

Yang, R. & Su, Z. (2010). Analyzing circadian expression data by harmonic regression based on autoregressive spectral estimation. *Bioinformatics*, 26, i168–i174.

Yerushalmi, S., Yakir, E. & Green, R.M. (2011). Circadian clocks and adaptation in *Arabidopsis*. *Molecular Ecology*, 20, 1155–1165.

Yon, F., Seo, P.J., Ryu, J.Y., Park, C.M., Baldwin, I.T. & Kim, S.G. (2012). Identification and characterization of circadian clock genes in a native tobacco, *Nicotiana attenuata*. *BMC Plant Biology*, 12, 172.

Zavala, J.A., Patankar, A.G., Gase, K., Hui, D. & Baldwin, I.T. (2004). Manipulation of Endogenous Trypsin Proteinase Inhibitor Production in *Nicotiana attenuata* Demonstrates Their Function as Antiherbivore Defenses. *Plant Physiology*, 134, 1181–1190.

Zeilinger, M.N., Farré, E.M., Taylor, S.R., Kay, S. a & Doyle, F.J. (2006). A novel computational model of the circadian clock in *Arabidopsis* that incorporates PRR7 and PRR9. *Molecular Systems Biology*, 2, 58.

Acknowledgements

On first place I want to thanks my parents, Yvonne and Carlos, who through the years taught me the appreciation for the natural things, especially the joy of a garden full of plants and the scientific curiosity for the natural phenomena that surrounds us.

Here, I want to thanks Prof. Ian Baldwin for the provided opportunity to do my Ph.D. research on this field that combines several scientific branches in one single place, and the scientific guidance he provided through these years. Not less, my gratitude to my group leader, Dr. Sang-Gyu Kim, for his guidance, fruitful talks and discussion, and the patience to understand each other scientific approaches during these years. My thanks to Dr. Tamara Krügel for her scientific knowledge on plants but also on gardening, and the opportunity for always be able to ask her advice. My gratitude to Prof. Maria Mittag for the insides on the chronobiology field.

Nonetheless, my thanks to the circadian group members who made the work much easier, and provided invaluable help when it was needed. Special thanks to Eva Rothe, whose support made possible that several of our plants kept working on time like a good clock, and taught me quite much of our analytical platform. In the same manner, I am thankful for the support of Dr. Danny Kessler, who supplied me with a lot of the natural knowledge of our field plot and of *Nicotiana* interactions. Also my thanks to the gardeners for taking care of the thousands of plants at glasshouse and chambers, sometime with even unusual growing conditions. Not forgetting the countless persons who answered so many of my questions through all this time, helped, pointed out and/or advised on the how to of several things.

To all my extant friends, still here or not (no one extinct thankfully), who through their friendship make my life in Jena quite amenable, full of nice shared moments at work, at the field and at wherever else we spend some time.

And my special thanks to my girlfriend, Francesca, not only for her support and patience, but also for her language skills so many times necessary to disentangle the length of my sentences.

I want morebooks!

Buy your books fast and straightforward online - at one of the world's fastest growing online book stores! Environmentally sound due to Print-on-Demand technologies.

Buy your books online at
www.get-morebooks.com

Kaufen Sie Ihre Bücher schnell und unkompliziert online – auf einer der am schnellsten wachsenden Buchhandelsplattformen weltweit! Dank Print-On-Demand umwelt- und ressourcenschonend produziert.

Bücher schneller online kaufen
www.morebooks.de

VDM Verlagsservicegesellschaft mbH
Heinrich-Böcking-Str. 6-8
D - 66121 Saarbrücken   Telefax: +49 681 93 81 567-9

info@vdm-vsg.de
www.vdm-vsg.de

Printed by Books on Demand GmbH, Norderstedt / Germany